Level I

Teacher's Manual

Rebecca W. Keller, Ph.D.

Cover design: David Keller
Opening page: David Keller, Rebecca W. Keller, Ph.D.
Illustrations: Rebecca W. Keller, Ph.D.

Copyright © 2004, 2007, 2010, 2011 Gravitas Publications, Inc.

All rights reserved. No part of this publication may be reproduced, stored in a retrieval system, or transmitted, in any form or by any means, electronic, mechanical, photocopying, recording, or otherwise, without prior written permission from the publisher. This publication may be photocopied without permission from the publisher only if the copies are to be used for teaching purposes within a family.

Level I Biology Teacher's Manual

ISBN 10: 1-936114-39-9
ISBN 13: 978-1-936114-39-9

Published by Gravitas Publications, Inc.
4116 Jackie Road SE, Suite 101
Rio Rancho, NM 87124
www.gravitaspublications.com

Printed in United States

A Note from the Author

This curriculum is designed to give students both solid science information and hands-on experimentation. Level I is geared toward fourth through sixth grades, but much of the information in the text is very different from what is taught at this grade level in other textbooks. I feel that students beginning with the fourth grade can grasp most of the concepts presented here. This is a *real* science text, so scientific terms are used throughout. It is not important at this time for the students to master the terminology, but it *is* important that they be exposed to the real terms used to describe science.

Each chapter has two parts: a reading part in the textbook and an experimental part in the laboratory workbook. In the teacher's manual, an estimate is given for the time needed to complete each chapter. It is not important that both the reading portion and the experimental portion be concluded in a single sitting. It may be better to split these into two separate days, depending on the interest level of the child, and the energy level of the teacher. Also, questions not addressed in the teacher's manual may arise, and extra time may be required to investigate these questions before proceeding with the experimental section.

Each experiment is a *real* science experiment and not just a demonstration. These are designed to engage the students in an actual scientific investigation. The experiments are simple, but they are written the way real scientists actually perform experiments in the laboratory. With this foundation, it is my hope that the students will eventually begin to think of their own experiments and test their own ideas scientifically.

Enjoy!

Rebecca W. Keller, Ph.D.

How To Use This Manual

The Biology Level I Teacher's Manual provides directions for the experiments in the Laboratory Workbook and answers to the questions asked in each experiment. Additional information for each chapter is provided as supplementary material in case questions arise while reading the text. It is not necessary for the students to learn this additional material as most of it is beyond the scope of this level. However, the teacher may find it useful for answering questions.

The laboratory section (or Experiments) together with the Review are found at the end of each chapter in this manual. All of the experiments have been tested, but it is not unusual for an experiment to fail. Usually repeating an experiment helps both student and teacher see where an error may have been made. However, not all repeated experiments work either. Do not worry if an experiment fails. Encourage the student to troubleshoot and investigate possible errors.

Getting Started

The easiest way to follow this curriculum is to have all of the materials needed for each lesson ready before you begin. A small shelf or cupboard or even a plastic bin can be dedicated to holding most of the necessary chemicals and equipment. The following *Materials at a Glance* chart lists all of the materials needed for each experiment. A materials list is also provided at the beginning of each lesson.

Several experiments require living organism. Some of these can be found in the backyard or local environment, or they can be purchased from a local pet store. All of the living things required for these experiments can be purchased from internet sources.

BIOLOGY LEVEL I | v
Materials at a Glance

Materials at a Glance

Experiment 1	Experiment 2	Experiment 3	Experiment 4	Experiment 5
Items such as: 　rubber balls 　cotton ball 　orange 　banana 　apple 　paper 　sticks 　leaves 　rocks 　grass 　Legos or 　　building 　　blocks, etc.	pencil or pen colored pencils 　or crayons	plant with at 　least 6 flat 　green leaves lightweight 　cardboard or 　construction 　paper tape 2 small jars marking pen	2 or more small 　jars 2 or more fresh 　white carnation 　flowers food coloring knife	2 small jars several pinto 　beans absorbent white 　paper plastic wrap clear tape 2 rubber bands marking pen knife

Experiment 6	Experiment 7	Experiment 8	Experiment 9	Experiment 10
microscope with 　a 10X objective[1] microscope 　slides[2] 3 eye droppers fresh pond 　water or 　water mixed 　with soil protozoa study 　kit[3] methyl cellulose[4]	microscope 　with a 10X 　objective[1] microscope 　slides[2] 3 eye droppers protozoa study 　kit from 　Experiment 6[3] baker's yeast distilled water Eosin Y stain[5]	caterpillars 　collected 　locally OR butterfly kit[6] small cage	tadpoles[7] tadpole food[7] small aquarium tap water 　conditioner and 　tap water OR distilled 　water	clear glass or 　plastic tank 　with a solid lid water plastic wrap soil small plants small bugs: 　worms 　small beetles 　ants, etc.

[1] Student Microscope, www.gravitaspublications.com, "Other Products" section, Gravitas Publications, Inc.
The following materials are available from Home Science Tools, www.hometrainingtools.com:
[2] Glass Depression Slides, MS-SLIDCON or MS-SLIDC12
[3] Basic Protozoa Set, LD-PROBASC
[4] Methyl Cellulose, CH-METHCEL
[5] Eosin Y stain, CH-EOSIN
[6] Butterfly Garden, LM-BFLYGAR
　Butterfly Pupae, LM-BFLYCUL
[7] Grow-A-Frog Kit, LM-GROFROG

Most of the items from Home Science Tools
are included in their Real Science-4-Kids
Level 1 Science Kit, RS-KTLEV1

Laboratory Safety

Most of these experiments use household items. However, some items are poisonous. Extra care should be taken while working with all chemicals in this series of experiments. The following are some general laboratory precautions that should be applied to the home laboratory:

- Never put things in your mouth unless the experiment tells you to. This means that food items should not be eaten unless tasting or eating is part of the experiment.

- Use safety glasses while using glass objects or strong chemicals such as bleach.

- Wash hands after handling all chemicals.

- Use adult supervision while working with iodine or glassware and while conducting any step requiring a stove.

Contents

CHAPTER 1:	**LIVING CREATURES**	**1**
	Experiment 1: Putting Things in Order	7
	Review	11
CHAPTER 2:	**CELLS: THE BUILDING BLOCKS OF LIFE**	**12**
	Experiment 2: Inside the Cell	18
	Review	24
CHAPTER 3:	**PHOTOSYNTHESIS**	**25**
	Experiment 3: Take Away the Light	29
	Review	32
CHAPTER 4:	**PARTS OF A PLANT**	**33**
	Experiment 4: Colorful Flowers	37
	Review	40
CHAPTER 5:	**HOW A PLANT GROWS**	**41**
	Experiment 5: Which Way Is Down?	46
	Review	50
CHAPTER 6:	**PROTISTS I**	**51**
	Experiment 6: How Do They Move?	55
	Review	59
CHAPTER 7:	**PROTISTS II**	**60**
	Experiment 7: How Do They Eat?	63
	Review	67
CHAPTER 8:	**THE BUTTERFLY LIFE CYCLE**	**68**
	Experiment 8: From Caterpillar to Butterfly	72
	Review	76
CHAPTER 9:	**THE FROG LIFE CYCLE**	**77**
	Experiment 9: From Tadpole to Frog	80
	Review	85
CHAPTER 10:	**OUR BALANCED WORLD**	**86**
	Experiment 10: Making an Ecosystem	90
	Review	93

Chapter 1: Living Creatures

Overall Objectives   2

1.1   The science of life   2

1.2   Taxonomy   2

1.3   The kingdoms   3

1.4   Further classification   5

1.5   Naming living things   5

1.6   Summary   6

Experiment 1: Putting Things in Order   7

Conclusions   10

Review   11

Time Required

Text reading	1 hour
Experimental	30 minutes

Materials

Items such as:
- rubber ball
- cotton ball
- orange
- banana
- apple
- paper
- sticks
- leaves
- rocks
- Legos or building blocks, etc.

CHAPTER 1
Living Creatures

Overall Objectives

This chapter introduces the field of science, called biology, that is concerned with the study of living things. In this chapter students will learn how living creatures are categorized. Because living creatures have unique features, the students will discover that it is often difficult to make exact categories for all living things, and often these categories are changed or become outdated. New classification schemes are constantly being tried and rejected.

1.1 The science of life

Point out to the students that living things differ from non-living things because they are "alive" and eventually cease to be alive, or die. Have the students think about what being "alive" means. Have them come up with their own definition of "living." Point out that non-living things do not reproduce themselves, are not independently mobile, do not consume food or water, and do not die.

Explain to the students that there are different disciplines of biology to deal with different aspects of living things. Some are:

- *Molecular Biology*: the study of the molecules, proteins, and DNA that make up living things
- *Cell Biology*: the study of the whole cells of living things
- *Physiology*: studies involving whole animal systems
- *Genetics*: studies that are concerned with the genetic information encoded in DNA
- *Ecology*: the study of living things and their interaction with the environment
- *Botany*: the study of plants
- 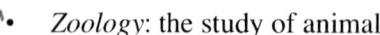 *Zoology*: the study of animals

1.2 Taxonomy

Taxonomy is that branch of science concerned with classifying living things. The term comes from the Greek word taxis meaning "arrangement" and nomos meaning "law." Carolus Linnaeus, a Swedish physician, began the systematic organization of living things into what is now called taxonomy. His system had only two kingdoms: Plants and Animals.

Have a discussion with the students about some of the many different kinds of living things — some are similar, and some are very different. Have the students list some animals that are similar, like wolves and domestic dogs, and some animals that are different, like jellyfish and humans.

1.3 The kingdoms

Before discussing the various taxonomic divisions, ask the students how they would classify living things. Ask them to make up their own groups. Answers will vary.

Because of the diversity of living things and because new information about known species continues to be discovered, it has been difficult to establish an overall consensus about how to classify all living things. In 1959, Cornell University Professor R.H. Whittaker proposed a five-kingdom system that is still used today and is the system that is presented in the textbook. However, a four-kingdom system that only recognizes the kingdoms Animalia, Plantae, Monera, and Virus has also been proposed. There is a proposed eight-kingdom system that recognizes the kingdoms Bacteria, Archaea, Archaezoa, Protista, Chromista, Plantae, Fungi, and Animalia. Also, a three-domain system has been proposed that would include the domains Bacteria, Archaea, and Eukarya.

Originally, the field of taxonomy was created to organize living things as a useful way to separate the various types of life into distinct categories for further study. Today, taxonomy is often a branch of evolutionary biology where an evolutionary connection is sought between certain species. For example, the distinction between the two domains Bacteria and Archaea is centered around differences in the ribosomal RNA sequences of species in these domains.

The first task in classifying living things is to determine into which kingdom a given organism should be placed. Before the discovery of the microscope, all known living things were classified as either plants or animals. They were placed in these two categories based on their plant-like or animal-like characteristics. However, today an organism is placed into a given kingdom based primarily on the cell type of that organism. Cells are discussed in more detail in Chapter 2.

Although there are five kingdoms, there are not five different cell types. There are two basic cells types — prokaryotic and eukaryotic (see Chapter 2). However, there are two different types of eukaryotic cells: plant cells and animal cells. A diagram showing the division of kingdoms based on cell type follows.

Eukaryotic					Prokaryotic
Plantae	Animalia	Fungi	Protista	Monera
plant cells	animal cells	like plant cells with some differences	like both plant and animal cells	

In spite of all of the classification difficulties, two main kingdoms stay consistent throughout all of the proposed classification schemes. These are the kingdoms, *Animalia* and *Plantae* (animals and plants).

The animal kingdom includes all animals. The common feature of all animals is the type of cell they have — the animal cell — which is covered in detail in the next chapter. This is not to say that all animal cells are the same. In fact, animal cells have a wide variety of shapes, sizes, and functions, but these cells all have some similar features that place all animals in the animal kingdom.

Have the students think about what creatures might be included in the kingdom *Animalia*. Then have them write down the names of several of these creatures. If it is not immediately obvious that an organism is an animal, have the students eliminate it from the other kingdoms first.

For example: A student may say "shrimp," but maybe it's not clear that a shrimp is an animal.

- Is it a plant? (Plantae) *No*.
- Is it a bacteria? (Monera) *No*.
- Is it a single-celled organism or does it have both plant and animal like characteristics? (Protista) *No*.
- Is it a fungus? (Fungi) *No*
- Then it must be an animal in the kingdom Animalia.

The same is true for all plants that are in the plant kingdom. All plants have plant cells, and again, this is covered in detail in the next chapter. Also, not all plant cells are identical: plant cells vary widely in shape, size, and function, but they all have similar basic features. Repeat the exercise above for plants. This may be trickier since a fungus can also look like a plant. Mushrooms, for example, are fungi, not plants. Use an encyclopedia or other resources if necessary.

The fungi originally were placed in the plant kingdom because they are sedentary like plants. However, fungi differ from plants in that they do not use the sun's energy for photosynthesis (see Chapter 3), and they differ from plants structurally. They are, therefore, in a kingdom of their own.

The fungi acquire their nutrients by absorption. They have special enzymes on the outside of their bodies that they use to break down food which they then absorb. There are three basic classifications of fungi: decomposers, parasites, and symbionts. The decomposers live off non-living organic matter such as fallen logs and animal corpses. The parasites absorb nutrients from a living host but can kill the host, and the symbionts acquire their food from a living host but are beneficial to the host they use.

The kingdoms Protista and Monera make up almost all of the microscopic organisms.

The cell type of Protista has characteristics of both plant and animal cells, but because Protista do not fit entirely in either the plant or the animal kingdom, they are given their own kingdom. These organisms are explained in more detail in Chapters 6 and 7.

The creatures in the kingdom Monera all have a particular cell type called a prokaryotic cell (see Chapter 2). Their cell shapes are diverse, but the most common shapes are rods, spheres, and spirals. *E. coli* is a type of bacteria that is rod-shaped and is found in our intestines. Bacterial pneumonia is commonly caused by a sphere-shaped bacterium.

1.4 Further classification

To further classify all living things, six additional categories are used. These differentiate between the various living things within a given kingdom.

The phylum, class, order, family, genus, and species are the names for these additional categories.

The phylum subdivides the kingdom into different groups. For the kingdom Animalia some of the phylum divisions are:

- Phylum Chordata—those that have a backbone
- Phylum Mollusca — includes clams
- Phylum Arthropoda — insects

The class further divides each phylum of a given kingdom. Some of the classes for the order Chordata are:

- Class Amphibia — frogs, toads, and newts
- Class Aves — birds
- Class Mammalia — animals with mammary glands for nursing
- Class Pisces — bony fishes

Then, the order divides the class, and the family divides the order.

1.5 Naming living things

The family is further divided into the genus and the species. These final classifications give each different creature a unique "two name" designation. The first name is the genus name. "Genus" comes from the Latin word for birth or origin and is the generic name for a given organism. It is written with a capital letter and is in italics. The second name, the species, is the specific name for that particular creature, and it is written in lower case and in italics.

The example shown in the text illustrates the different names for four different types of cats. It is not important that the students remember all of these names, only that they understand how each animal is classified.

Point out that the scientific names of all living things are in either Latin or Greek that has been "Latinized" by the addition of Latin endings. Linnaeus, although he spoke Swedish, used Latin to name living things. Latin was the universal language of scientists in his day, and he wrote most of his scientific work in Latin to make it available to other scholars. In general, the Latin or Latinized Greek name of a living thing often describes some unique feature of that organism. The name for the bobcat is *Felis rufa* which refers to its reddish coat. *Felis* is the Latin word for cat, and *rufa* in Latin means red, or ruddy.

Not all names given to a specific organism reflect some scientific aspect of that creature. The scientist that discovers the organism has the right to name it. Some names reflect where the organism was found, and other names may be derived from Greek myths or in honor of a person.

1.6 Summary

Review the summary points of this chapter with the students.

- Biology is the study of living things Review the difference between a living and a non-living thing. Discuss that taxonomy is a branch of biology that deals with classifying living things.

- The classification of living things begins with dividing them into various groups. Review the diagram with the students and point out the different groups. The largest groups are the kingdoms. Each kingdom is divided into phyla, each phylum into classes, each class into orders, each order into families, and each family into genera (plural form of genus) and species. The classification of a creature into a certain group depends on many things, like its cell type, whether it lays eggs, whether or not it has a backbone, etc.

- Point out to the students why classifying living things is important. By knowing how organisms are different or how they are similar, scientists can better understand how they live. For example, by observing the balancing behavior of a domestic cat on the edge of a narrow ledge, an understanding of how the mountain lion navigates the terrain of the Rockies may be possible. Also, observing the enmity between the neighborhood cat and the family dog may help explain the lack of cooperation between the lioness and the jackal.

Experiment 1: Putting Things in Order   Date: _____

Objective   In this experiment we will organize a variety of objects into categories.

Materials

Collect a variety of objects. Some suggestions are: rubber ball, cotton ball, orange, banana, apple, paper, sticks, leaves, rocks, grass, Legos or building blocks, etc.

Experiment

❶ Spread all of the objects out on a table. Carefully look at each object and note some of its characteristics. For example, some objects may be smooth, some fuzzy; some may be edible, others not; some may be large, some small, etc.

❷ Record your observations for each item in the Results section.

❸ Now try to define "categories" for the objects. For example, some objects may be "hard," so one category could be called "Hard." Some objects may be "round," so another category could be "Round." Try to think of at least 4 or 5 different categories for your objects. Write the categories along the top of the graph in the Results section.

❹ List the objects in the category that describes them. Take note of those objects that fit into more than one category. Write these objects down more than once, placing them in all of the categories that describe them.

❺ Next, take a look at each of the categories and each of the objects in those categories. Can you make "subcategories?" For example, some objects may all be the same color, so "Red" could be a subcategory. Some may be food items so "Food" could be a subcategory. Pick three categories and try to list several subcategories for each of these main categories.

❻ List the objects according to their category and subcategory.

BIOLOGY LEVEL I | 7
TEACHER'S MANUAL

In this experiment, the students will try to organize different objects according to some property like shape, color, or texture.

There are no "right" answers for this experiment, and the categories the students pick will vary.

Have the students collect a variety of objects, place the objects on a table and then make careful observations. Guide them to notice some features of the objects, such as color, shape, and texture. Also, discuss any common uses, for example, those used as toys or those used as writing instruments.

Have the students make some notes about the objects they have collected, briefly describe each object and list a few of its characteristics.

Next, have the students determine some overall categories into which the objects can be placed. For example, marbles, cotton balls, and oranges are round, so "Round" could be a category. Basketballs, baseballs, and footballs are all balls, so another category could be "Types of Balls." Some items may fit into more than one category. Basketballs can fit into both the category "Round" and the category "Types of Balls." Have the students write down each item in all of the categories where it fits.

Have the students look at each category separately and then choose three categories to further divide into subcategories. Guide them in thinking about what the subcategories might be.

8 | CHAPTER 1
Living Creatures

The students will now record the characteristics of the various items. Help them to be as descriptive as possible.

For example, oranges can be described as round, orange, sweet, food, living, etc. Tennis balls are round, fuzzy, yellow or green (or another color).

Help the students describe, in detail, several different items. Some examples are listed.

Results

(Answers will vary.)

Item	Characteristics
orange	round, orange, food, sweet
grape	oval, food, sweet, green
tennis ball	round, green, fuzzy
marshmallow	white, soft, sweet, shaped like a cylinder
cotton ball	round, white, fuzzy

BIOLOGY LEVEL I
TEACHER'S MANUAL

Categories					
white	*fuzzy*	*round*	*square*	*hard*	
marshmallow cotton ball	tennis ball cotton ball	oranges tennis ball cotton ball			

Categories	*White*		*Fuzzy*			
Sub-categories	*round*	*food*	*round*	*toys*		
	cotton ball	marshmallow	cotton ball	tennis ball		

Have the students write the categories at the top of each column using a PENCIL, so they are able to change the categories as more items are being written down.

They will decide which items fit into which categories and then write those items in the column below the category name.

Next, they will pick one to three of the categories and divide them further into subcategories, trying to choose categories that allow all of the items to ultimately be listed. If necessary, they can rename some of the main categories to better fit the items listed. The names of the categories and subcategories can be adjusted as needed so that each item is listed in a category and subcategory, but it's possible that not all of the items can be placed in a category and a subcategory.

This can be quite challenging. The point of this exercise is to illustrate the difficulty of trying to find a suitable organizational scheme for things with different characteristics.

CHAPTER 1
Living Creatures

Help the students write valid conclusions about the data they have collected. For example:

- *Both oranges and cotton balls are round.*
- *Both cotton balls and marshmallows are white.*
- *Tennis balls and cotton balls are both fuzzy.*

Examples of conclusions that are not valid:

- *Both cotton balls and marshmallows are white. Marshmallows are sweet so cotton balls are sweet.*
- *Tennis balls and cotton balls are both fuzzy. Tennis balls are bouncy so cotton balls must be bouncy.*

It is important to use only the data that has been collected and not make statements about the items that are not backed up by the data. It is obvious that marshmallows and cotton balls are both white, but it is not true that cotton balls are sweet. Because two or more items have one or two things in common does not mean that all things are common between them. Discuss this observation with the students.

Discuss the difference between valid and invalid conclusions. A valid conclusion is a statement that generalizes the results of the experiment, but draws only from the data collected. It does not go beyond the results of the data to include things that haven't been observed and does not connect results that should not be connected. An invalid conclusion is a statement that has not been proven by the data, or a statement that connects the data in ways that are not valid. The example

Conclusions

_____
_____
_____
_____
_____
_____
_____
_____
_____
_____
_____
_____
_____
_____
_____
_____
_____
_____
_____
_____
_____
_____
_____
_____
_____
_____

Review

What is taxonomy? _the branch of biology concerned with classifying living things_

List the five kingdoms. _Plantae_ _Animalia_
Fungi _Monera_
Protista

List the other six categories for classifying living things.

Phylum _Class_
Order _Family_
Genus _species_

Which kingdom are dogs, cats, and frogs in? _Animalia_

Which phylum are dogs, cats, and frogs in? _Chordata_

Which class are frogs in? _Amphibia_

Which order are dogs in? _Carnivora_

Which family are cats in? _Felidae_

What is the Latin name given to humans and what does it mean?

Homo sapiens. This means "man wise."

given is that marshmallows are sweet and white, but although cotton balls are also white, it is invalid to say they are sweet like marshmallows.

Chapter 2: Cells: The Building Blocks of Life

Overall Objectives	13
2.1 Introduction	13
2.2 The cell—a small factory	13
2.3 Types of cells	15
2.4 Prokaryotic cells	15
2.5 Plant cells	16
2.6 Animal cells	16
2.7 Organelles	16
2.8 Summary	17
Experiment 2: Inside the Cell	18
Conclusions	23
Review	24

Time Required

Text reading   1 hour
Experimental   1 hour

Materials

pencil or pen
colored pencils or crayons

Overall Objectives

In this chapter students will be introduced to cells and their proper names and structure. The most important points for them to remember are that cells are the fundamental building blocks of all living things and cells are highly complex and highly ordered.

2.1 Introduction

ALL living things are made of cells, and there are no living things that are not made of cells. This unifying concept was not fully realized until the middle of the 19th century. Theodor Schwann and Matthais Jakob Schleiden, in separate publications in 1838 and 1839, presented the cell doctrine — the theory that all living things are composed of small units that we call cells.

Today, we know much more about what cells look like, what they contain, and how they work. However, it is important to point out to the students that, although we know much more about cells than the early cell biologists of the 19th century did, we are still very far from fully understanding cells in their entirety. Even the simplest cells are far more complex than ever imagined.

Cells, like non-living things, are ultimately composed of atoms. The atoms combine to form molecules. Some of the molecules that are found in living things are also found in non-living things. But many molecules in living things are found ONLY in living things. DNA and proteins are examples.

Some organisms are composed of only one cell. But many organisms are made of lots of cells. These multi-cellular organisms are often composed of tissues that then form organs that fit together to form the whole organism.

2.2 The cell—a small factory

The cell is like a small factory. There are many different molecules working together in a very systematic and orderly fashion. Even the simplest cells, the prokaryotic cells, that lack some of the features found in eukaryotic cells, are highly organized (see Sections 2.4 and 2.5).

The illustration in this section of the student text shows some of the activities common to all cells.

- Proteins and small molecules are moved in and out of cells.
- Large molecules are manufactured from smaller molecules inside certain areas of a cell.
- Molecules are transported from place to place within the cell.
- Some molecules are stored for later use.

Point out to the students how the activities inside cells are coupled to each other. Ask the students to think about how a city works. Have them list some of the jobs that various people do in the city; for example, postal workers carry the mail, sanitation personnel pick up the trash, etc. Now have the students think of just one job; for instance, bringing milk to the grocery store. Have them think about all of the people and all of the tasks that must be done to get milk to the store. The cows need to be milked by the farmer, the milk needs to be processed by a dairy, the milk cartons need to be made by the factory, the milk cartons need to be delivered to the dairy, the dairy workers need to fill the milk cartons with milk, etc. Now ask the students what would happen if even one of those people did not do their job — could the job get finished? Explain to the students that this is similar to how things work inside a cell.

The illustration on page 13 of the student text is a "key" for the cell factory illustration on page 12. Nucleic acids actually look like those in the drawing, but the large and small molecules are just symbols rather than drawings of actual molecules.

Page 13 of the student text explains some of the activities that take place in cells and describes the molecules that are involved in these activities.

Proteins do most of the work in cells. These are the cell's main machinery, and they are responsible for transporting molecules within the cell as well as to and from separate cells. Proteins called enzymes are responsible for all the manufacturing of molecules, such as making other proteins and nucleic acids or making energy molecules.

The nucleic acids are responsible for carrying the cell's "library." These molecules carry the information the cell needs in order to make new cells or other proteins in the cell.

All of the molecules in the cell work together to keep the cell functioning. Every molecule has a particular job. Also, the cell knows how many molecules it has working at a particular time. Sometimes more molecules are made and sometimes fewer of the same molecules are made. This depends largely on what the cell is doing and the surrounding environment. As the environment changes, the cell adjusts to these changes and may alter the manufacturing of certain molecules.

Point out to the students that, unlike non-living things, cells die. It is not true that the whole of a cell can be explained by the sum of its parts because, even though all of the parts may still be around, cells die. A dead cell cannot come back to "life." Living cells can only come from other living cells and not simply from the machinery that composes them.

2.3 Types of cells

There are two basic cell types: prokaryotic cells and eukaryotic cells. All organisms are made of one of these two cell types.

Prokaryotes are bacteria and are in the kingdom Monera. Prokaryotes lack a central membrane-bound nucleus that holds the genetic material (DNA).

Eukaryotes do have a nucleus. Plants, animals, fungi, and protists are all composed of eukaryotic cells.

Eukaryotic cells are generally larger than prokaryotic cells with the total volume of a eukaryotic cell often about a thousand times larger than that of prokaryote.

2.4 Prokaryotic cells

Prokaryotes are considered simple cells because they lack many features that eukaryotes possess.

For example:

- Prokaryotes do not have a membrane-bound nucleus, while eukaryotes do.

- Prokaryotes have a single DNA chromosome, and it is circular. Eukaryotes have several linear chromosomes.

- The inside of the cell (cytoplasm) of a prokaryote is almost devoid of structure. There is no cytoskeleton (no microtubules or actin as found in eukaryotes), and there are no organelles, which eukaryotes do have — see Section 2.7.

- Energy production in prokaryotes is a function of the whole cell and not the sole function of an organelle, as it is in eukaryotes.

Many prokaryotic cells have flagella or pili that move the organism. A flagellum is a long whip that is attached to a sophisticated motor that twirls the whip at great speed.

A cell uses pili to attach itself to surfaces and to other cells. The pili are actually much longer than they are shown in the illustration in the textbook.

There are about 20,000 distinct bacteria that are known. In general, bacteria are tough, with some having the ability to withstand extreme heat or extreme cold. Some bacteria are harmless and even beneficial to humans. Lactobacillus acidophilus is a common bacterium found in the digestive tract, and it aids in the digestion of food. Other bacteria can cause disease in humans and even death; for example, Salmonella enterica is a bacterium that can cause food poisoning.

2.5 Plant cells

Plants are made of eukaryotic cells. Plant cells have both a membrane-bound nucleus and organelles.

Have the students carefully examine the illustration in this section of the student text. It is not important that they remember all of the names of the organelles. Functions of some of the organelles are listed in Section 2.7. Have the students refer to this chart as they examine the cell.

Point out that plant cells have a cell wall. The cell wall gives plants the stiffness they need for standing upright. Plants also have chloroplasts. Chloroplasts are organelles that are used by the cell to make food by photosynthesis (see Chapter 3). These two features, cell walls and chloroplasts, make plant cells different from animal cells.

2.6 Animal cells

Animal cells have many of the same organelles as plant cells. Both plant and animal cells have mitochondria, a nucleus, ribosomes, and a cytoskeleton. The cytoskeleton is a network of microtubules and actin (a protein) that is used for maintaining structural features of the cell and for transporting molecules from place to place within the cell.

As we have already seen, animal cells differ from plant cells in several important ways. First, animal cells do not have chloroplasts and do not use the sun's energy to make food. Animal cells also do not have cell walls, and they lack the central vacuole found in plant cells. So, although plant and animals cells have many similar features and are both eukaryotes, plant and animal cells are also different.

Point out to the students that the drawing in the student text represents a generalized cell and that there are many different kinds of animal cells. We have bone cells, nerve cells, and skin cells, all with slightly different features. ALL of these cells are eukaryotic cells, but each is specialized to perform a particular task for our bodies.

2.7 Organelles

The table in the student text shows functions for some of the organelles in eukaryotic cells. It is not important that the students memorize these functions, only that they understand that cells are highly organized and eukaryotes have organelles that perform particular tasks. Many organelles are membrane-bound; that is, they have a plasma membrane on the outside.

2.8 Summary

Discuss the main points of this chapter with the students.

Have the students try to think of a living thing that isn't made of cells. Review that all living things are made of cells.

Discuss with the students the complexity of cells and how they function like small factories. Look at the illustration on page 12 of the student text, and discuss briefly the various jobs that go on inside the cell. Ask the students what would happen to the cell if one or more of the jobs wasn't performed. For example, what would happen if the pores would not let molecules out (molecules would build up in the cell), or what would happen if no energy molecules were made (the cell would run out of energy), and so on.

Remind the students that eukaryotic cells have organelles, and prokaryotic cells *do not*. This does not mean that prokaryotic cells are not complex, only that they do not have the same features as eukaryotic cells.

Ask the students to think about various organisms and what kind of organs they have. Then ask them what the organs are made of. Discuss with the students how each organ is important for the organism and performs some function.

For example: A cow has a stomach and a stomach is made of stomach tissue. A stomach is used for digesting food. A cow cannot live without a stomach.

Point out that some organs are not essential for life (eyes) but that they are needed for a special purpose (seeing).

CHAPTER 2
Cells

In this exercise the students will examine the similarities and differences between various cell types.

All cells share some common features. One such feature is DNA (deoxyribonucleic acid). DNA is often referred to as the genetic code. Almost every cell has DNA. The DNA in a cell contains many volumes of information that the cells need to make proteins, metabolize nutrients, grow, and divide.

Another feature common to all cells is that they have ribosomes which make proteins from RNA (ribonucleic acid). RNA is different from DNA but is still a nucleic acid. RNA is made from DNA and proteins are made from RNA.

DNA->RNA->proteins

In living cells there are no known exceptions to this paradigm. Proteins are always made from RNA, and the RNA used to make proteins is always made from DNA.

Experiment 2: Inside the Cell   Date:_____

Objective Using the drawings of the three types of cells in the student text, we will observe which features of the three different types of cells are similar and which are different.

Materials

pen or pencil
colored pencils or crayons

Experiment

(Some answers may vary.)

Write down some things you observe in the drawings that are similar for all three cell types:

All cells contain DNA.

All cells contain ribosomes.

All cells have something that holds them together, like a cell wall or plasma membrane.

Write down some observations of things that are different:

Prokaryotic cells do not have a nucleus.

Animal cells do not have a cell wall.

Plant cells contain chloroplasts, but animal cells do not.

Write down the function of each of the following:

Nucleus — *in eukaryotic cells, holds together the DNA and the proteins needed to use the DNA*

Mitochondria — *organelles that make energy; found in plant and animal cells*

Chloroplasts — *organelles that use the sun's energy to make food; found in plant cells*

Cell wall — *stiff outer membrane found in plant cells that makes the plant sturdy*

Lysosome — *the place where big molecules get broken down*

Peroxisome — *the place where poisons in the cell are removed*

How do animal cells differ from plant cells, and how do both plant and animal cells differ from bacterial cells?

List as many reasons as you can for the differences between bacteria, plants, and animals and why you think their cells may differ.

Bacteria (have or don't have...)	**Plants** (have or don't have...)	**Animals** (have or don't have...)

BIOLOGY LEVEL I | 19
TEACHER'S MANUAL

Have the students look up bacteria in the encyclopedia.

Some facts about bacteria:

- They may be spherical, rod shaped, or spiral.
- They live in many different environments including soil, water, organic matter, and in the bodies of plants and animals.
- They are autotrophic (make their own food)
- or saprophytic (live on decaying matter)
- or parasitic (live off a live host).
- They can be either beneficial or harmful to humans.

Plants have organs, as do animals, and therefore need to be made of many different types of cells. Have the students think about the different parts of a plant, such as the leaves and roots, and discuss what the cells of each might need to do. (For example: Root cells need to take up minerals from the soil. Since roots are in the dirt, they do not have to be green like leaves. The leaves are green because they need to use chloroplasts for collecting light.) Also, discuss how plant cells differ from animal cells. For example, plants don't have

20 | CHAPTER 2
| Cells

bones, and they don't usually move, so they don't need muscles like some animals do. Have the students think of a variety of animals like deer, fish, and frogs, and then discuss the differences between them. Next, have them write down why there are different types of cells in these different creatures.

Have the students fill in the blanks in the drawing on this page. Have them first try to do it without looking at the student text. Have them color the organelles. The colors do not need to match those in the textbook.

As the students fill in the blanks, discuss the functions of the various parts of the cell. Point out how the structures differ and, where possible, point out how the structure of the part matches its function. For example, the flagellum looks like a whip and is used like a whip for swimming. And, the cell membrane and cell wall are used to enclose the contents of the cell, and so they are thin and extend around the outside of the cell..

Without looking at your text, fill in the blanks with as many names for the structures in the cell as you can. Color the cell.

Is this an animal cell, a plant cell, or a prokaryotic cell? Write the cell type at the top.

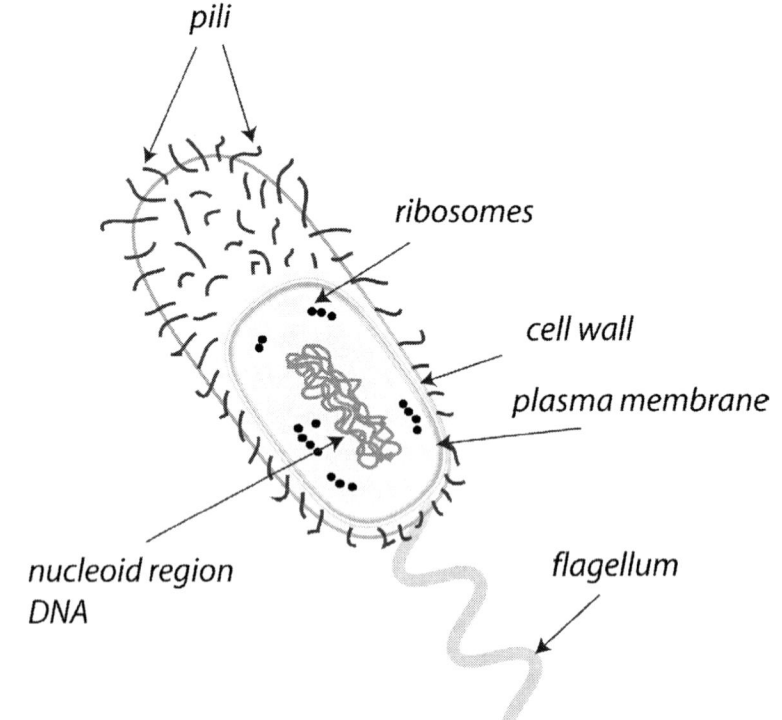

Without looking at your text, try to fill in as many of the structures as you can.

Is this an animal cell, a plant cell, or a prokaryotic cell? Write the cell type at the top.

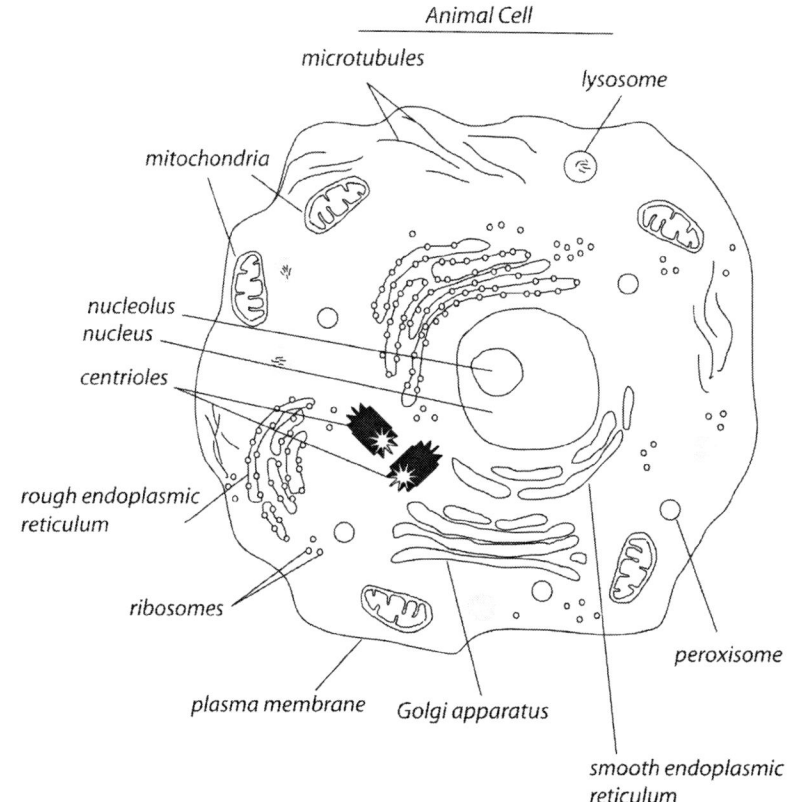

Have the students fill in the blanks in the drawing on this page. Have them first try to do it without looking at the text. Have them color the organelles. The colors do not need to match those in the text.

22 | CHAPTER 2
| *Cells*

Have the students fill in the blanks for the drawing on this page. Have them first try to do it without looking at the text. Have them color the organelles. The colors do not need to match those in the text.

Without looking at your text, fill in the blanks with as many names for the structures in the cell as you can. Color the cell.

Is this an animal cell, a plant cell, or a prokaryotic cell? Write the cell type at the top.

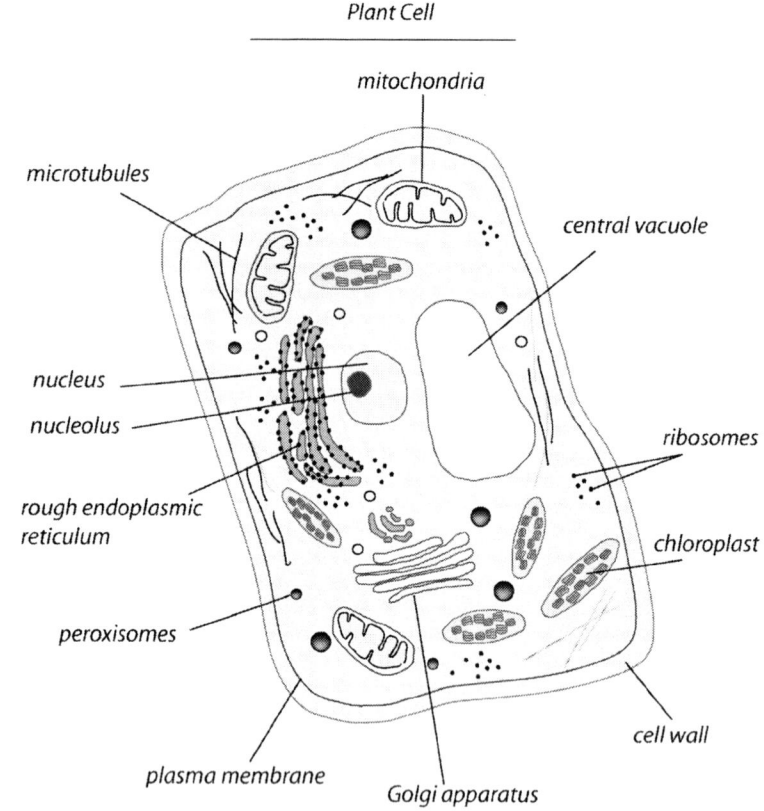

Plant Cell

Conclusions

Help the student arrive at some conclusions about cells based on what they have learned in this chapter.

Some examples include:

- *All living things are made of cells.*
- *All living things have DNA.*
- *Not all cells are alike.*

CHAPTER 2
Cells

Review

(Some answers may vary.)

Answer the following questions:

What are cells made of?   *Cells are made of a highly organized arrangement of atoms and molecules.*

What are tissues made of?   *Tissues are made of cells.*

What are organs made of?   *Organs are made of tissues.*

Define the following terms:

prokaryote   *the simplest type of cell — a bacterium is a prokaryote*

eukaryote   *a cell that has a nucleus — plants and animals are eukaryotes*

organelle   *a little organ inside a cell that performs a certain task*

Name some organelles found in plant and animal cells:

chloroplast

mitochondria

nucleus

peroxisome

golgi apparatus

lysosome

What does a flagellum do?   *A flagellum is used by some organisms for movement.*

Where are chloroplasts mostly found?   *Chloroplasts are mostly found in plants.*

What are mitochondria for?   *Mitochondria are used by the cell to make energy molecules.*

Chapter 3: Photosynthesis

Overall Objectives	26
3.1 Introduction	26
3.2 Chloroplasts	26
3.3 Why plants have leaves	27
3.4 Photosynthesis in other organisms	27
3.5 Summary	28
Experiment 3: Take Away the Light	29
Conclusions	31
Review	32

Time Required

    Text reading   1 hour
    Experimental   1 hour

Materials

    plant with at least 6 flat, green leaves
    lightweight cardboard or construction paper
    tape
    2 small jars
    marking pen

Overall Objectives

In this chapter the students will be introduced to photosynthesis, the process by which plants harvest the sun's energy to make food.

3.1 Introduction

Ask the students how they think plants eat. Have them look at the picture in the student text, and ask them why plants can't eat cheeseburgers. Answers will vary, but try to get them to understand that plants can't eat cheeseburgers because plants don't have:

- *teeth*
- *mouth*
- *stomach*
- *saliva*
- *fast food restaurants*

Explain that plants are photosynthetic, i.e., they use the sun's energy to make their own food. Not all plants are solely photosynthetic. There are a few carnivorous and parasitic plants that obtain additional nutrients by non-photosynthetic means.

For example, mistletoes and dodders are parasitic plants that obtain nutrients from other photosynthetic, food-making plants. These plants wrap long vines around the stems of photosynthetic plants and penetrate the stems with tiny root-like appendages. In this way, they obtain water, sugar, and other nutrients from the host.

There are over 600 carnivorous plants that have been identified. The best-known examples are the Venus flytrap, sundews, and pitcher plants. These plants obtain additional nutrients by capturing insects. They can also use photosynthesis and survive without capturing insects but grow faster and are greener when they are supplied with insects.

3.2 Chloroplasts

Photosynthesis occurs in specialized organelles called chloroplasts. The actual biochemical steps are quite complex and beyond the scope of this level. However, a schematic outline is presented below which shows the essential steps.

The overall reaction during photosynthesis is:

carbon dioxide (CO_2) + water (H_2O) + light = sugar + oxygen (O_2)

The leaves of photosynthetic plants contain small holes (pores) called stomata that can open and close, allowing carbon dioxide to enter and oxygen to exit.

Chlorophyll molecules "pick up" the light energy and transfer this energy to other molecules in a series of reactions that ultimately produce sugar molecules. This is an example of "light energy" being converted into "chemical energy." The details of this process are not important at this level of study.

The chlorophyll molecules are located in the thylakoid, a membrane-bound compartment in the chloroplast. Stacked thylakoids form the granum.

It is hard to overemphasize the importance of photosynthesis to all living things. All creatures need a source of energy or they will die. Plants use photosynthesis to make their food, and they get this energy from the sun. We, and all other animals, get our energy by eating plants or by eating other animals that eat plants. The sun's energy is ultimately the source of energy for all living things. Plants put this energy into a form usable by animals, including ourselves.

3.3 Why plants have leaves

Leaves are the main organ for carrying out photosynthesis in plants. Many leaves are broad in order to help them collect as much light as possible.

Discuss with the students how a broad leaf will pick up lots of light, and a small, narrow leaf will pick up less light (the bigger and broader the leaf, the more exposure to the sun it has). If possible, have the students collect a few varieties of leaves from their yard or neighborhood, and discuss with them the various shapes they find. Also, show the students how the leaves can pick up sunlight from all directions by being attached to the tree at different angles. You can do this by simply having the students look at a tree and try to find out which leaves have direct sunlight hitting them. Then, later in the day, look at the tree again to see that different leaves are now in direct sunlight.

Conifers are evergreen trees that have a different type of leaf. The leaves on these trees are narrow and thin. These narrow leaves, called needles, allow conifers to live in areas with little water. The needles have a thick outer coating, called a cuticle, that reduces water loss. Although the needles collect less sunlight, their shape and structure make them able to retain more water.

3.4 Photosynthesis in other organisms

Land plants, in the kingdom Plantae, are not the only organisms that carry out photosynthesis. Seaweed and microscopic algae are usually classified in the kingdom Protista and also use photosynthesis to make food. However, some of the multicellular varieties of algae, like red algae, are still placed in the plant kingdom, or in the "red plant" kingdom.

Many algae have chlorophyll that has the same molecular structure as that of land plants and, in addition, have chlorophylls of slightly different molecular form. These additional chlorophylls give algae their characteristic red or brown coloration.

Other organisms that use photosynthesis to make food include the cyanobacteria and several different protists (see Chapter 6).

Cyanobacteria were once considered algae, but are now known to be prokaryotes and are classified in the kingdom Monera. Like green land plants, cyanobacteria contain chlorophyll and have thylakoids. They can exist as single organisms or in colonies. "Pond scum" is typically made of several different varieties of cyanobacteria.

3.5 Summary

Discuss with the students the main points from this chapter.

- Ask the students how plants get their food.
  most by photosynthesis, a few as parasites, and some by eating insects

- Ask the students to describe photosynthesis in as much detail as they can. Have them look at the picture of the chloroplast in Section 3.2 of the student text.

- Ask the students to explain which parts of a plant participate in photosynthesis.

stems	*yes*
flower buds	*yes*
fruit	*yes, some*
leaves	*yes*
roots	*no*

Experiment 3: Take Away the Light   Date: _____

Objective *In this experiment we will remove the sunlight from several plant leaves and observe any changes that occur over a period of many days. We will then compare the changes in these leaves to the changes in several other leaves that have been removed from the plant but exposed to light and supplied with water. This will allow us to observe if light and water are sufficient to keep the plant leaves alive.*

Hypothesis _____
_____
_____

Materials

- plant with at least 6 flat, green leaves
- lightweight cardboard or construction paper
- tape
- 2 small jars
- marking pen

Experiment

1. Take some cardboard or construction paper and cut it into squares large enough to completely cover a leaf. Make enough pieces to cover the front and back of 3 leaves.

2. Six different leaves will be tested. Two of the leaves will be left on the plant (attached) and four leaves will be removed from the plant (unattached).

3. Using the cut out cardboard pieces from Step 1, tape two pieces so they cover the front and back of one of the two attached leaves. The other leaf will stay uncovered. The four unattached leaves will be either covered and placed in water, uncovered in water, uncovered out of water, or covered out of water.

4. With the marking pen, label the leaves in the following manner:

   Leaf 1:   UA — uncovered, attached

   Leaf 2:   CA — covered, attached

   Leaf 3:   UUW — uncovered, unattached, in water

   Leaf 4:   CUW — covered, unattached, in water

   Leaf 5:   UU — uncovered, unattached (no water)

   Leaf 6:   CU — covered, unattached (no water)

BIOLOGY LEVEL I | 29
TEACHER'S MANUAL

In this experiment the students will examine the effects of removing light from the leaves of a photosynthetic plant.

Have the students first read the experiment through to determine what is being investigated. Then have them write the objective.

Next, discuss the possible outcomes for each of the leaves. Ask the students the following questions before they write the hypothesis:

- *What will happen to Leaf 1? It is uncovered and attached to the plant. It can get sunlight and also can get water and other nutrients from the rest of the plant.*

- *What do you think will happen to Leaf 2? It is covered and attached to the plant. This means it cannot get any sunlight, but it does get water and other nutrients from the rest of the plant.*

- *What will happen to Leaf 3? It is uncovered (will get sunlight), not attached to the plant (no additional nutrients), but it is placed in water, so it will receive water.*

- *What will happen to Leaf 4? It is covered (will not get sunlight), is unattached (will not get additional nutrients), but it will get water.*

- *What will happen to Leaf 5? It is uncovered (will get sunlight), but will get no water or nutrients.*

- *What will happen to Leaf 6? It is covered (will get no sunlight), and it will not*

30 | CHAPTER 3
Photosynthesis

get any water or additional nutrients.

Have the students guess which leaves will die first, which will not die at all, and which may survive a short time. Help them write a suitable hypothesis based on this discussion. Some examples are:

- *All of the leaves not in water will die.*
- *Only the leaves without sunlight will die.*
- *Only leaf 6 will die.*

The results for this experiment may vary depending on the type of plant that is used. Tree leaves can also be used.

Two of the leaves used in this experiment are "controls." The leaf that is attached and uncovered is a positive control (Leaf 1). This leaf should remain healthy throughout the course of the experiment unless something happens to the plant. The leaf that is covered and unattached and without water is a negative control (Leaf 6). This leaf should die first.

Discuss the use of controls with the students. Positive controls help the investigator determine if the experimental setup is working. Negative controls tell the investigator when the desired effects of the experiment have indeed occurred. Both types of controls are useful for making sure the experimental results are valid. In other words, if the positive control did not work (the leaf dies on the plant), then when the other leaves die it cannot be concluded that the leaves died because of the changes made during the experiment. Likewise, the negative control should be a

❺ Take two small jars and fill them with water. Take the two leaves that will be placed in water and prop one in each jar, keeping the stems submerged. Check the water level every day over the course of the experiment to make sure there is enough water in the jars.

❻ Wait several days, and then observe the changes to the leaves daily by carefully removing the cardboard and then re-taping it. Record your observations.

Results

	UA	CA	UUW	CUW	UU	CU
Day 1						
Day 2						
Day 3						
Day 4						
Day 5						
Day 6						
Day 7						
Day 8						
Day 9						
Day 10						

Conclusions

negative result (the leaf should die without water or sunlight) to prove that the other samples survive or die accordingly. If the negative control survives, it is an indication that something else is happening — the leaves are especially tough, the experiment needs more time, a plastic plant was used instead of a real one and so the leaves cannot die (anything is possible). Positive and negative controls give the experimenter the boundaries of the experiment and allow him/her to make valid conclusions about the samples in between.

Have the students record the results in the boxes. Have them write down any observation they see. Short one or two word descriptions like, "green," "green with some brown," "mostly brown," "wrinkled," are fine.

Have the students write valid conclusions based on their results. Help them to be accurate.

For example:

- *Leaf 1 survived for one week.*

- *Leaf 2 did not survive past two days.*

- *Leaf 3 survived for one week. Leaf 3 needs only sunlight and water to live one week unattached from the plant.*

- *All of the leaves without sunlight turned yellow-brown, but did not die.*

Have the students discuss what they learned. Is it true that water and sunlight alone are sufficient for the survival of a detached leaf? How about for a leaf that remains attached, but lacks sunlight — is sunlight required for this leaf to survive?

CHAPTER 3
Photosynthesis

Review

Define the following terms:

photosynthesis — *Photo means "light" and synthesis means "to make." Photosynthesis is the process used by plants to make food using sunlight.*

chloroplast — *A chloroplast is a tiny organelle inside the leaf of plants. It is where photosynthesis occurs.*

chlorophyll — *Chlorophyll is a molecule inside a chloroplast that "captures" the sun's energy for photosynthesis.*

conifer — *A conifer is a type of evergreen tree. It has thin needlelike leaves and makes cones.*

algae — *protozoa that use photosynthesis to make food like land plants do*

cyanobacteria — *bacteria that use photosynthesis to make food*

How do most green plants get their food? Do they eat cheeseburgers?

Most land plants get their food using photosynthesis. They never eat cheeseburgers.

Chapter 4: Parts of a Plant

Overall Objectives	34
4.1 Introduction	34
4.2 How plants live	34
4.3 Parts of a plant: roots	35
4.4 Parts of a plant: stems	35
4.5 Parts of a plant: flowers	36
4.6 Summary	36
Experiment 4: Colorful Flowers	37
Conclusions	39
Review	40

Time Required

Text reading 1 hour
Experimental 1 hours

Materials

2 or more small jars
2 or more fresh white carnation flowers
food coloring
knife

Overall Objectives

In this chapter the students will learn about the features that make up the majority of land plants. By the end of this lesson, the students should be able to list the parts of flowering plants.

4.1 Introduction

Plants are the main food producers on Earth. There are more than 250,000 different land plants known with an additional 150,000 plants not found on land. Plants grow in most areas except where there is extreme heat or cold; for example, at both the north and south poles and the driest parts of the desert. Discuss with the students the importance of plants and how plants provide food for other living things. Have them think of what they eat, and then have them try to think of something they eat that doesn't come directly or indirectly from plants. For example, they may say "marshmallows" as something they eat that's not from a plant. However, marshmallows are made mostly of sugar and sugar comes from sugar cane — a plant!

Have a discussion with the students about the many different kinds of plants we use for food. Ask them to identify which parts of plants they think we eat. Is a cherry a flower? Is an asparagus a leaf? Is an ear of corn a root?

Here are some examples:

- Cherries, apples, pears, oranges, avocados — fruit
- Lettuce, cabbage, spinach, kale — leaves
- Celery — stems and leaves
- Asparagus, broccoli — stems and buds
- Artichokes — flower buds
- Peas, corn, beans — seeds
- Carrots, turnips, red beets — roots

Have a discussion with the students about how we don't eat all types of plants and that some plants are even poisonous.

4.2 How plants live

Most plants are designed to live partly in soil and partly in the air.

Plants have specialized tissues that allow them to obtain nutrients from the earth's soil, from sunlight, and from the earth's atmosphere. As we saw in Chapter 3, the leaves of a land plant are able to "capture" the sun's energy to make food. Leaves are also designed to exchange oxygen for carbon dioxide from the surrounding air. The roots of a plant are designed to take in nitrogen, minerals, and water from the soil.

There are two main parts common to land plants: the shoot system and the root system. The shoot system exists above the soil and consists of the stems, leaves, and, in flowering plants, the flowers. The root system is that part of the plant that exists below the surface of the soil.

The components that make up these systems are called organs. The organs of plants are made of many cells that perform particular tasks. For example, different types of cells do different jobs in the vascular tissue of vascular plants (those having xylem and phloem — see Section 4.4). Other cells form the epidermis, which covers the outer areas; some form the cortex, which is just inside the epidermis; and on the leaves, there are guard cells that open and close to allow air and moisture in.

All of the organs work together to maintain the life of the plant.

The stem supports the leaves and flowers, transports water and minerals from the roots, and even stores food. The roots anchor the plant to the soil, absorb water and minerals from the soil, and store food. The leaves make food with photosynthesis and export the food to the rest of the plant. When any one of these organs is cut off from the rest of the plant, it does not easily survive on its own.

4.3 Parts of a plant: roots

The root system serves two main functions:

- to anchor the plant to the soil
- to absorb nutrients from the soil

Some roots, especially the roots of plants like sugar beets, carrots, and sweet potatoes, are also food storage organs for the plant.

There are two main types of root systems: the taproot system and the fibrous root system.

The taproot system has a central root that goes straight down, deep into the soil. Smaller roots branch off from this central root and are called feeder roots. The taproots of many desert plants extend several meters into the soil to reach deep sources of water.

Fibrous roots are different from taproots in that there is no central root. Instead, several main roots extend downward with smaller roots extending from them. These form a dense mass. Grasses have this kind of root system, and the many small roots help secure grasses firmly to the soil.

Roots have special tissues that allow the absorption of minerals and water from the soil. They have a large surface area which aids this absorption, most of which occurs at the tip of the root where the outer surface is thin and where there are many root hairs (small roots branching out from the larger roots). The root hairs contain most of the surface area of roots.

4.4 Parts of a plant: stems

The shoot system of a plant consists of the stems, leaves, and flowers.

The stems of a plant are usually long and contain nodes and buds. Some stems are vertical and stand erect. Some are creeping stems, and some are vines that climb other plants or solid structures.

Stems contain two main tissues: xylem and phloem. These two tissues are responsible for transporting food, minerals, water, and other nutrients from the leaves to the roots and from the roots to the leaves.

The xylem and phloem can be arranged in concentric rings surrounding the core tissue which is called the pith. The xylem and phloem can also be intermixed, still surrounding the pith. This varies depending on the type of plant. The pith can also extend pith rays which radiate from the central core. Food storage is the principle function of the pith.

The xylem transports water and minerals from the roots to the leaves at a rate of about 15 meters per hour for most plants. The fluid in the xylem rises against gravity by capillary action without the use of molecular motors or pumps. The capillary action is created by evaporation of water from the leaves. As water evaporates from the leaves, a negative tension or pressure pulls the water up from below.

The transport of sugars in the phloem works by a different mechanism than the transport of minerals in the xylem. The phloem sap moves downward through a series of sieve tubes connected to each other end to end. The sugar is transported with small protein pumps that pump it into the phloem tissue.

4.5 Parts of a plant: flowers

Flowers are the reproductive organs of flowering plants.

Flowers form from the shoot and contain four floral organs called the sepals, petals, stamens, and carpels. Flowers that contain all four floral organs are called complete flowers. Some flowers, like those on most grasses, lack petals and are referred to as incomplete flowers.

The stamen is the "male" part of the flower and the carpel is the "female" part. At the tip of the stamen is the anther which releases pollen grains. Pollen is collected at the tip of the carpel and travels down the pollen tube into the base where the eggs are housed inside the ovary. The pollen fertilizes the egg, giving rise to an embryo that then develops into a seed.

After the pollen grains fertilize the eggs inside the ovary, the flower dies. A fruit develops, holding one or more seeds depending on the species. When the ripe fruit falls off the plant, it can be carried to other places by the wind or animals. If the conditions are right, a seed will grow into a new plant.

4.6 Summary

Discuss with the students the summary statements listed in this section of the student text.

Experiment 4: Colorful Flowers Date: _____

Objective *In this experiment we will observe the flow of water from the base of a stem to the flower.*

Hypothesis _____

_____

Materials

- 2 or more small jars
- 2 or more fresh white carnation flowers
- food coloring
- knife

Experiment

1. Take two or more of the small jars, and add water and several drops of food coloring.

2. Trim the end of one carnation stem, and place it in one of the jars of colored water.

3. Watch the petals of the carnation, and record any color changes observed.

4. Take the carnation out of the jar, and cut a small slice of the stem off the bottom. Try to identify the xylem and the phloem. In the Results section, draw a picture of what you see. Cut the carnation flower lengthwise. Try to identify the parts of the flower.

5. Take one stem and slice it lengthwise with a knife, starting about halfway up the stem and cutting away from the flower. (Have an adult help you.) Stick one end of the divided stem into a solution of colored water, and place the other part of the stem in a jar that contains water of a different color. Let the carnation soak up the colored water until the petals begin to change color. In the Results section, draw a picture of what you observe.

BIOLOGY LEVEL I | 37
TEACHER'S MANUAL

In this experiment the students will observe the transport of water through the xylem of carnation stems. Have the students read the experiment in its entirety before writing the objective. An example is shown. Answers may vary.

Two variations of transport will be investigated. First, simple transport up the stem of the carnation will be examined. This part of the experiment serves to establish a control so that predictions about the second variation can be made. The transport observed in the first variation should result in petals with single colors. This experiment demonstrates that colored water is indeed transported up the stem to the petals of the flower.

In the second variation, the stem is split lengthwise starting about halfway up the stem and cutting away from the flower. Each of the divided stem ends is placed in a separate jar with a different color of water.

Have the students predict what will happen in each case. Ask the following questions:

- *Will the colored water travel to the petals of the flower?*

- *Will two different colors travel up to the flower in the second experiment?*

- *Will the petals on the flower in the second experiment be a single color, or will the petals have two colors?*

- *Do you think the colored water will travel straight up in the second experiment, or will the colors mix in the stem? How could you tell which is happening?*

CHAPTER 4
Parts of a Plant

Have the students write a hypothesis based on their answers to these questions. For example:

- *The colored water will not travel to the flower.*
- *The split stem will not allow any colored water to travel to the flower.*
- *The split stem will give two colors in the petals.*

Answers may vary.

Have the students record their results. Petals can be removed from the flower and taped into the workbook in the Results section. The petals will not be uniformly colored, but will have a dark strip of color at the very end. If enough time is allowed, the veins in the petals will also turn colors.

The split carnation will result in a flower with two colors — one color on each side of the flower. The colored water travels up the split stem and colors only the petals on the same side as the piece of stem. The colors do not mix, and the split stem does not prevent the colored water from being transported.

VARIATIONS (optional)

If more carnations are available, some additional questions can be investigated. For example:

- *What happens if a carnation is allowed to soak in one color and is then transferred to a different color? Do the colors blend at the tip of the petals, or are the petals striped?*

Results

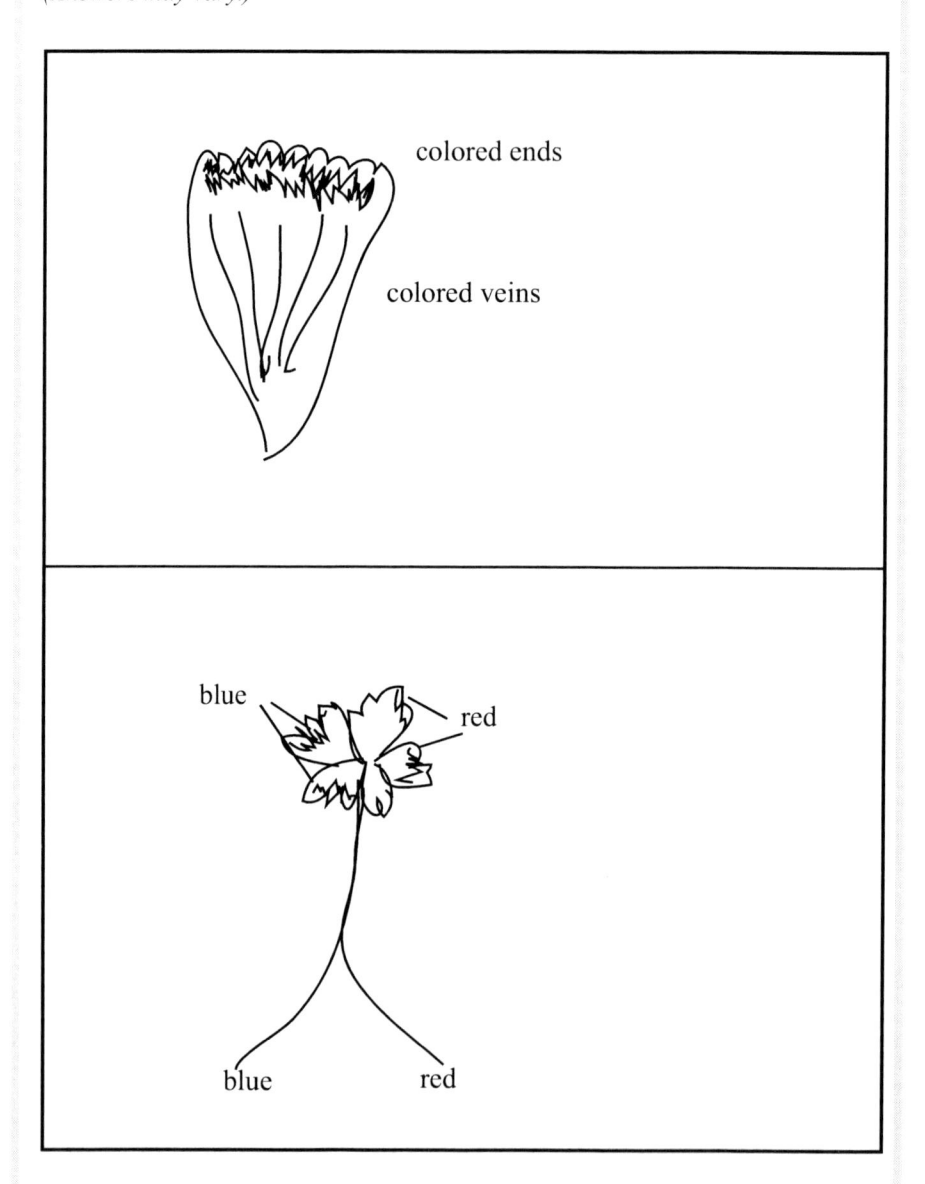

(Answers may vary.)

colored ends

colored veins

blue   red

blue   red

Conclusions

- *What happens if one carnation is left for a short period of time without any water, and another is placed in water for the same amount of time. When the first carnation is then placed in water, will one "drink" water faster than the other?*

- *What happens if some of the petals are removed from the flower? Will the remaining petals be darker in color or color faster? Put two flowers in the same colored water solution. Remove several petals from one flower and compare the petal colors of each flower after they have soaked for a time.*

CHAPTER 4
Parts of a Plant

Review

Define the following terms:

- **xylem** — *a tissue found in the stems of plants that is used to move water and minerals up from the roots*
- **phloem** — *a tissue found in the stems of plants that is used to move food down the plant to the roots*
- **pith** — *the tissue used for food storage that is found in the center of a stem*
- **pollen** — *small grains used by flowers to make seeds*
- **stamen** — *the part of the flower that holds the place where the pollen is made*
- **ovary** — *the base of the carpel where the egg is located*

Name the four main parts of a flowering plant:

- *stems*
- *roots*
- *leaves*
- *flowers*

Name two types of roots:

- *fibrous roots*
- *taproots*

Chapter 5: How a Plant Grows

Overall Objectives	42
5.1 Introduction	42
5.2 Flowers, fruits and seeds	42
5.3 The seed	43
5.4 The seedling	43
5.5 Signals for plant growth	43
5.6 Plant nutrition	44
5.7 The life cycle of flowering plants	44
5.8 Summary	45
Experiment 5: Which Way is Down?	46
Conclusions	49
Review	50

Time Required

 Text reading   1 hour
 Experimental   1 hour

Materials

 2 small jars
 several dried beans (pinto, navy, etc.)
 absorbent white paper
 plastic wrap
 clear tape
 2 rubber bands
 knife

Overall Objectives

In this chapter the students will examine how plants grow from seeds to full-sized plants, with the focus being on the life cycle of flowering plants. Only flowering plants have seeds, while other plants reproduce with spores.

5.1 Introduction

Have a discussion with the students about how plants grow. Ask them if they can think of a plant that does not start out as a seed. Explain to the students that although not every plant has flowers, every flowering plant starts off as a seed. Although not covered in the student text, you may wish to explain that non-flowering plants reproduce from spores that are made up of one or more cells. Spores have a tough outer coat, are more primitive than seeds, have little stored food (in contrast to seeds), and are adapted to be easily dispersed and to survive for extended periods of time. Ferns and mosses are examples of non-flowering plants that reproduce with spores.

5.2 Flowers, fruits, and seeds

All flowering plants begin life as tiny seeds that are formed inside the flower. There are many different kinds of flowers. Some are very small and are difficult to see with the unaided eye, like the flowers of duckweed. Some flowers are very large, like the Rafflesia, which grows on the floor of dark tropical forests and gets as large as 3 to 4 feet in diameter.

Following fertilization with pollen, the flower produces a fruit at its base. The fruit protects the seed and may help in the germination of the seed. Simple fruits contain one ovary at the end of one carpel (see Chapter 4, Section 4.5) and produce one seed. The avocado is an example of a simple fruit containing one seed. Aggregate fruits, such as strawberries, blackberries, and raspberries, have flowers that each contain many carpels and many ovaries, resulting in many seeds bunched together in one fruit. Multiple fruits differ slightly from aggregate fruits and are formed by the individual ovaries of several flowers which bunch together to make one fruit. A pineapple is an example of a multiple fruit.

Seeds come in different shapes and sizes. Some seeds, like the maple seed and dandelion seed, are designed to be carried by the wind and have propeller shaped fruit or tufts of hairs that help them float in the wind. Other seeds, like those that are enclosed in burrs, are designed to be carried away by animals. Have the students think about various seeds and their shapes and discuss whether the seeds are carried away by wind or animals.

5.3 The seed

Before the seed is released, a small embryo or baby plant develops. It is housed inside the seed and gets food from the cotyledons (kä-tə-lē′-dənz) which are the fleshy part of the seed. Both the embryo and the cotyledons are encased in a tough outer coating called the seed coat that protects the seed from harm. Seed coats can be smooth, like that of a bean, or rough and sculptured, like that of a peach or plum seed.

5.4 The seedling

When the conditions are right, including the presence of adequate moisture, the seed will begin to germinate. The first indication of germination is the swelling of the seed as it absorbs water. Then, a small root called the radicle emerges. The radicle grows downward into the soil, ensuring that the germinating seed will have enough water and nutrients for full growth. For seeds such as beans, peas, and onions, a small hook is next formed from the radicle which helps the seed push through the soil. The new stem then straightens and, as sunlight hits the seedling, new leaves emerge and begin making food by photosynthesis. These new leaves are called foliage leaves.

Not all seeds germinate in exactly the same way. A wheat grain, for example, does not form a hook for pushing through the soil, but instead, the stem emerges as a straight shoot.

The cotyledons are the food reserves for the seed. Once photosynthesis has begun, the cotyledons are no longer needed. They wither and fall away from the new seedling.

5.5 Signals for plant growth

There are many different chemical signals that tell plants how to grow. In general, these signals are called tropisms. A tropism is anything that causes a plant to turn toward a stimulus. Tropism comes from the Greek word *tropos* which means "turn."

There are three main tropisms that tell plants how to grow:

- phototropism
- gravitropism
- thigmotropism

Photo means "light," so phototropism means "turning toward light." This is the signal plants get from sunlight, and plants will always grow toward sunlight.

Gravitropism means "turning toward gravity." Roots exhibit positive gravitropism, and shoots exhibit negative gravitropism. Roots grow in the direction of gravity, and shoots grow opposite the direction of gravity.

Thigmo means "touch," so thigmotropism means "turning toward touch." Climbing vines exhibit this type of tropism. A vine will grow straight until it contacts another surface. This contact stimulates the vine to coil around the surface.

Point out the various types of stimuli that cause plants to grow. Have a discussion with the students about how these are "signals" much like road signs, stop signs and traffic lights that tell drivers where to turn and where not to turn, when to go and when to stop.

5.6 Plant nutrition

Although plants make all of their own food with photosynthesis, they require certain mineral nutrients for proper growth. Mineral nutrients are absorbed by the roots from the soil.

There are 17 essential mineral nutrients that are required for healthy plant growth. These include phosphorus, calcium, magnesium, potassium, sulfur, nitrogen, iron, boron, manganese, zinc, copper, and nickel, among others. Some mineral nutrients are required in large amounts, like nitrogen and calcium, but others are only needed in small quantities, like zinc and copper.

Mineral deficiencies in plants are usually obvious by the change in leaf color or overall growth of the plant. A tomato plant that is deficient in magnesium, for example, will have yellow leaves.

Mineral nutrients are involved in many of the biochemical mechanisms of the plant. For example, magnesium is needed for chlorophyll production, and calcium is required for maintaining membrane structure. Phosphorus is needed for cell membranes and for making DNA, and nitrogen is needed for making proteins and DNA.

5.7 The life cycle of flowering plants

The figure in this section of the student text summarizes the life cycle of flowering plants.

The students should be familiar with each of the steps of the life cycle: flower to seed-containing fruit, fruit to germinating seed, seed to seedling, seedling to young plant, and finally, young plant to mature flowering plant.

It should again be pointed out to the students that there are other plants that do not make seeds or have flowers but reproduce by other means.

For example, ferns and horsetails are seedless plants that reproduce using spores. Mosses are a different kind of seedless plant that also use spores for reproduction.

New plants can also be grown from existing plants. "Cuttings" can be removed from some plants and, when placed in water, new roots will grow. Some plants also put out thin extended stems which will sprout a new baby plant and roots. So, other mechanisms of reproduction besides seed formation from flowers can be found in the plant kingdom.

5.8 Summary

Have a discussion with the students about the main points to remember which are listed in this section of the student text.

CHAPTER 5
How a plant grows

In this experiment the students will investigate the signals of gravitropism and phototropism.

Have the students read the entire experiment before writing the objective and hypothesis. An example objective is shown.

Before writing the hypothesis, have the students predict whether or not the direction in which the seed is placed in the jar will affect the growth of the roots and stems. Also have them predict what effect the lack of sunlight will have on those beans that are placed in the dark.

Example hypotheses:

- *Only the beans facing downward (bean D) will grow properly.*
- *Only beans A and B will grow properly.*
- *The direction of the bean will not make any difference, and all beans will grow with the roots downward and the stems upward.*
- *The stems for the beans in the dark will not grow upward.*

Any type of dried bean should work for this experiment. Peas will also work, but the orientation of the unsprouted peas may be more difficult to determine.

Two jars will be used. One jar will be placed in the dark, and the other jar will be kept exposed to sunlight. The question that will be addressed is whether or not sunlight is responsible for making the shoots grow upward, or whether gravity is enough to cause the roots to grow downward and the shoots to grow upward even without sunlight.

Experiment 5: Which Way Is Down?   Date: _____

Objective *In this experiment we will observe the growth of several bean seeds. We will examine the direction the roots and stems grow in.*

Hypothesis _____
_____
_____

Materials

2 small jars
several dried beans (pinto, navy, etc.)
absorbent white paper
plastic wrap
clear tape
2 rubber bands
knife

Experiment

❶ Cut strips of white paper to fit around the inside of the jars.

❷ Label each strip "A", "B", "C", and "D" with a few centimeters between each letter.

A   B   C   D

❸ Place the beans in different directions on the labels, and fasten them with clear tape. They should look like this:

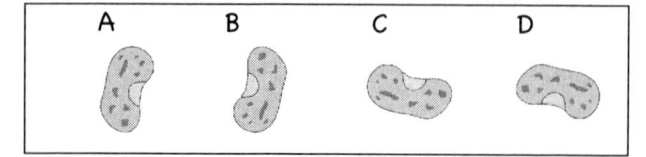

A   B   C   D

❹ Place the paper with the attached beans gently inside the small jar. The beans should be between the jar and the paper.

❺ Place one or two more beans in between the paper and the jar, but don't tape these. These beans will be opened and examined before the roots completely emerge.

❻ Add some water to the bottom of the jar, but don't submerge the beans.

❼ Cover the jar with plastic wrap and secure it with a rubber band. Place the jar in direct sunlight.

❽ Using the second jar, repeat steps 1-7, but place the jar in a dark place like a closet.

BIOLOGY LEVEL I | 47
TEACHER'S MANUAL

❾ In a few days the beans will start to grow. When the beans begin to change, take out one of the loose beans and gently cut it open. Try to identify the different parts of the embryo.

❿ Continue to observe the growth of the beans. Watch and record their changes every few days. Try to determine if the beans placed in different directions grow differently, and compare the beans grown in the light with those grown in the dark.

Results

Draw the parts of the embryo in the box below.

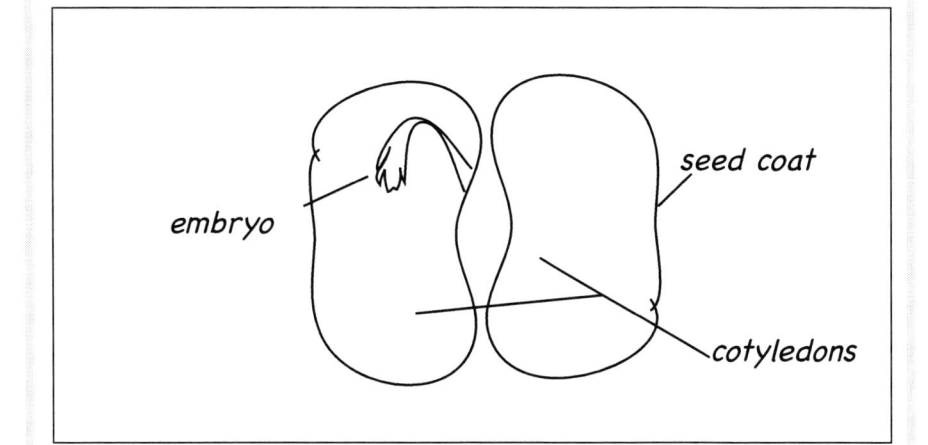

embryo, seed coat, cotyledons

Sometimes, one or more of the beans will fail to grow. If this happens, any beans that didn't grow can be replaced with new ones. Soak the new beans in water overnight and replace any failed bean with a water-soaked one. Beans may rot if soaked for more than a few days.

Place only a little bit of water in each jar and secure the top with plastic wrap. Allow the paper to contact the water, but do not allow the beans to soak in the water. It may be necessary to add more water to the jars as the experiment proceeds.

Before any visible shoots begin to emerge from the beans, have the students take out one or more of the loose beans. Help them carefully cut the beans lengthwise. A small embryo should be visible. Have the students draw what they observe. Have a discussion with them about how the fleshy part of the bean provides food for the embryo during the early stages of germination.

CHAPTER 5
How a plant grows

Have the students record the changes in the beans as they grow. It should take about two weeks for the beans to have good roots and long shoots. Make sure to continue the experiment until all of the beans are growing. Any beans that fail to grow can be replaced with the extra beans.

The radicle should begin to emerge from the bean around Day 5, although this may vary depending on the amount of water and light present. Initially, the roots grow in the direction in which the bean is placed; some will grow up, some will grow sideways and some will grow down. By Day 10, the roots should turn and begin to migrate downward for all of the beans.

Have the students compare the beans in the light with those in the dark. The direction of the roots should be independent of the presence of light, and all of the beans should have roots extending downward.

Have the students note any differences between each of the beans.

Record your observations for each bean. It may be several days before you see a change. Record the number of the day that you observe a change, and draw a picture of the way each bean looks on that day.

(Drawings may vary.)

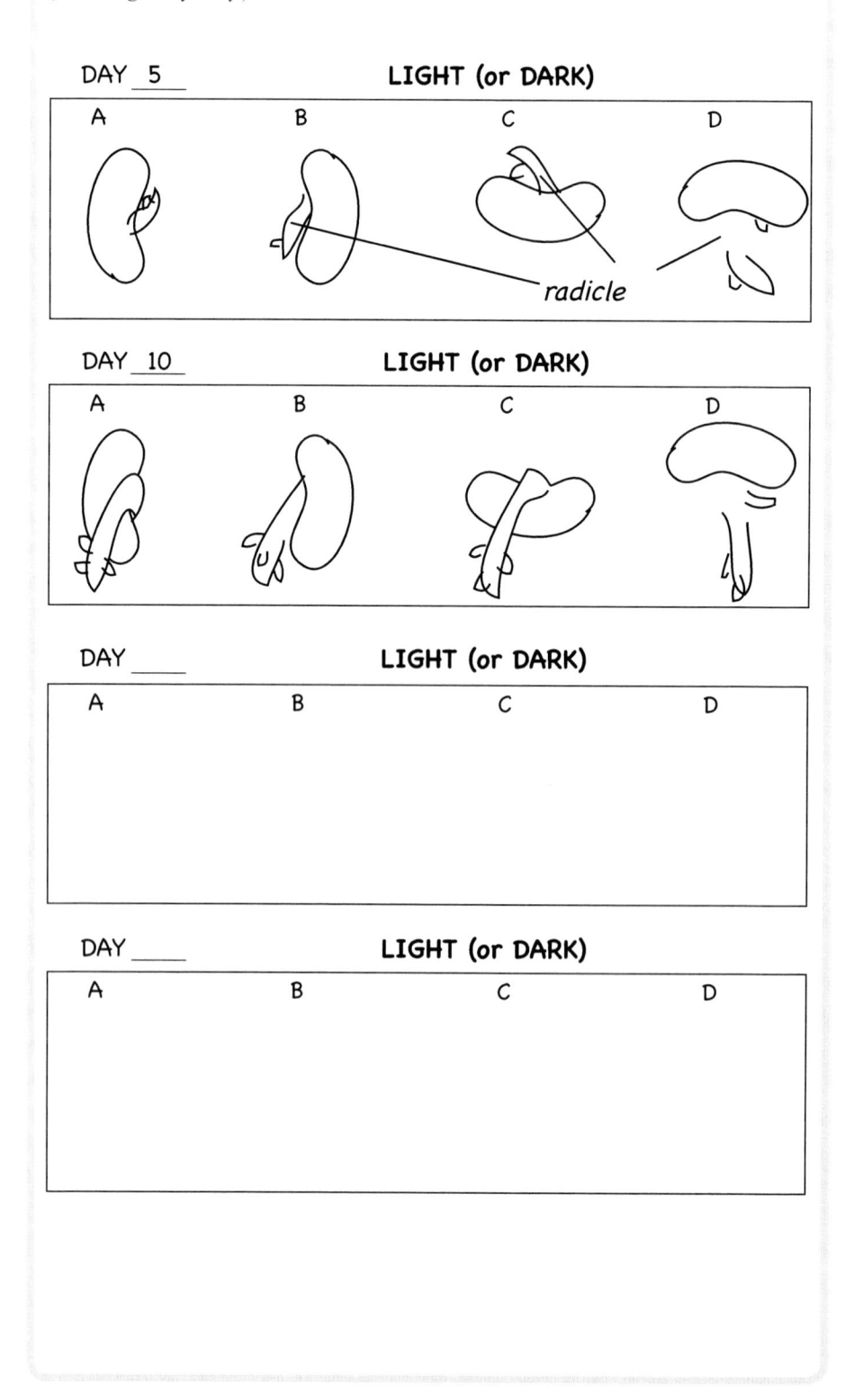

DAY __5__   LIGHT (or DARK)

A   B   C   D

radicle

DAY __10__   LIGHT (or DARK)

A   B   C   D

DAY _____   LIGHT (or DARK)

A   B   C   D

DAY _____   LIGHT (or DARK)

A   B   C   D

Conclusions

Have the students write valid conclusions based on the data they collected.

Some example are:

- *The beans in the dark did not grow differently than the beans in the light.*
- *The direction in which the beans were placed did not affect the direction in which the roots and stems grew.*
- *None of the beans grew.*

It is important that the conclusions the students write are based only on the data collected for this experiment even if the experiment did not work.

Review

(Answers may vary.)

Define the following terms:

seed	*a small plant embryo and supply of food encased in a tough outer shell*
seedling	*the young plant that forms from a seed*
seed coat	*the tough outer shell of a seed*
cotyledon	*the fleshy part of the seed that supplies the embryo with food*
embryo	*the small plant inside a seed*
germination	*the process of a seed beginning to grow and turning into a plant*

Name two plant signals.

sunlight

gravity

List the four stages in the life cycle of a flowering plant.

flower to fruit containing seeds

seeds to germinating plant

germinating plant to young seedling

young seedling to mature plant

Chapter 6: Protists I

Overall Objectives	52
6.1 Introduction	52
6.2 The microscope	52
6.3 Movement	53
6.4 Summary	54
Experiment 6: How Do They Move?	55
Conclusions	58
Review	59

Time Required

Text reading	1 hour
Experimental	1 hour

Materials

microscope with a 10X objective [1]
microscope slides [2]
3 eye droppers
fresh pond water or water mixed with soil
protozoa study kit [3]
Protoslo [4]

[1] Student Microscope, www.gravitaspublications.com, "Other Products" section, Gravitas Publications, Inc.
The following materials are available from Home Science Tools, www.hometrainingtools.com:
[2] Glass Depression Slides, MS-SLIDCON or MS-SLIDC12
[3] Basic Protozoa Set, LD-PROBASC
[4] Methyl Cellulose, CH-METHCEL

Overall Objectives

The students will be introduced to the microscopic organisms known as protists. They will also be introduced to the function and use of a microscope.

6.1 Introduction

Protists are members of the kingdom Protista.

[NOTE: In some texts, the term protist refers only to the microscopic species. In the 1970's and 80's, the boundaries of the kingdom Protista were expanded to include some multicellular organisms, such as seaweeds and slime molds, and the name of the kingdom was changed to Protoctista. In this chapter we will focus only on the microscopic varieties which will be called protists or protozoa.]

Protists are found wherever there is water, including saltwater, freshwater, and soil. Some well-known protists include:

- malarial parasites
- red tide organisms
- diatoms
- potato blight organisms
- Giardia
- African sleeping sickness organisms

Protists have been difficult to classify because they are eukaryotes and can have both plant-like and animal-like qualities. Protist is an "umbrella term" that fits those organisms that cannot be easily placed into any other kingdom.

6.2 The microscope

This section introduces the students to the light microscope. There are many different types of microscopes in use today, including light, electron, scanning tunneling, and scanning force microscopes.

Light microscopes use lenses and light to magnify small objects and make them appear larger. If a light microscope is available, it would be helpful to discuss the various parts of the microscope at this time.

When using a light microscope, a sample is placed below the lens, and a light source is placed below the sample. The observer looks through the lens at the sample below. If the lenses are powerful enough, individual cells can be seen. Plant and animal cells are between 10 and 100 microns in size (one micron is 1/1000 of a millimeter). A human hair is about 200 microns in diameter. Bacteria are smaller than plant or animal cells at about 1 micron in diameter. The unaided eye can easily visualize objects as small as 1 mm so a magnification of 10X to 1000X allows us to see cells the size of animal cells as well as those of bacteria.

Robert Hooke is traditionally credited with observing the first cells in 1665. However, Galileo adapted lenses for use in microscopy as early as 1614. Hooke seems to have coined the term "cell," which comes from the Latin word *cella* which means small room or cubicle. It appears that Hooke did not observe living cells. Anton van Leeuwenhoek is credited with the discovery of microscopic animals that he called "animalcules." While studying pond water, he observed the organisms we now call protists.

6.3 Movement

In this section of the student text the three main groupings of protists are presented. These groupings are based primarily on how the organism moves.

The three groups of protists are:

- ciliates
- flagellates
- amoebas

The ciliates include Paramecium and Stentor. The members of this group have bodies that contain many small hair-like projections, called cilia. The cilia "beat" rhythmically, propelling the organism smoothly through the water. By controlling the cilia's beating pattern and speed, ciliates can turn and even go backwards.

The flagellates include the uniflagellate (having one flagellum) species like Euglena, and the dinoflagellates (having two flagella) like Ceratium and Pfiesteria. Flagellates move by using whips that propel them through water.

Although they have a simple appearance in a microscope, cilia and flagella are actually very sophisticated machines. Each whip contains strands of long aggregate molecules called microtubules. As the microtubules slide past each other, the flagellum or cilium changes orientation. When the microtubules slide past each other in the opposite directions, the whip again changes orientation. These successive changes cause the cilia or flagella to beat or whip causing the protozoan to move.

The amoebas (also called rhizopods) do not have flagella or cilia to use for swimming, but rather use pseudopodia to move and feed. Pseudopods are "false feet" that allow the organism to creep along surfaces.

Amoebas are found in both freshwater and saltwater environments. Some amoebas are harmful to humans, such as Entamoeba histolytica, which causes amoebic dysentery.

6.4 Summary

Discuss the main points of this chapter with the students.

Ask the students to think about the plant-like and animal-like qualities that protists possess. Ask them how protists eat and move. Have them identify whether they eat and move like either plants or animals. For example:

- *Some protists use the sun's energy to make food just like plants do.*
- *Some protists eat other protists just like animals sometimes eat other animals.*
- *Amoebas use "false feet" to move and crawl like animals.*

Ask the students to explain how a microscope makes things look bigger, and have a discussion about how some things are too small for our eyes to detect, and so a microscope must be used in order to see them. Ask the students to think about how exciting it must have been for the first microscopists to discover this new world of tiny organisms.

Have a discussion with the students about the different types of movement that protists display and the way in which protists are characterized by their method of movement.

Experiment 6: How Do They Move?   Date _____

Objective *In this experiment, three types of protozoa will be observed. Protozoa found in pond water will be characterized based on how they move.*

Hypothesis *We can tell the difference between ciliates and flagellates in pond water.*

We can only tell the difference between ciliates, flagellates and amoebas in pond water.

Materials

- microscope with a 10X objective
- microscope slides
- 3 eye droppers
- fresh pond water or water mixed with soil
- protozoa study kit
- methyl cellulose

Experiment

❶ Familiarize yourself with your microscope before beginning this experiment. Read the instruction manual for your microscope, if it is available, and try to look at any prepared samples that may have come with your microscope. If you already know how to operate a microscope, skip this step.

❷ Take one of the samples from the protozoa kit, and place a small droplet onto a glass slide that has been correctly positioned in the microscope.

❸ Observe the movement of the protozoa. If the organisms move too quickly, apply a droplet of methyl cellulose to the glass slide.

❹ Patiently observe the movement of one type of protozoan. Note the type of protozoan you are observing in the Results section. Try to describe how the protozoan moves. Draw the protozoan, and write down as many observations about its movement as you can.

❺ Repeat Steps 2-4 with the other two protozoan types.

❻ Now take a droplet of fresh pond water (or water mixed with soil) and place it in the microscope. Try to determine the type of protozoan you are observing based on how the organism moves. Write and draw your observations in the Results section.

BIOLOGY LEVEL I | 55
TEACHER'S MANUAL

[NOTE: This is an optional experiment. If a microscope is unavailable, or if protozoa cannot be ordered, this experiment can be skipped.]

In this experiment the students will examine the three different types of protozoa discussed in this chapter of the student text, and then they will examine pond water to identify individual protozoa based on their method of movement.

Have the students read the experiment in its entirety before writing an objective and a hypothesis. Examples are given.

Microscope recommendation: Gravitas Publications [www.gravitaspublications.com] sells an inexpensive "student" microscope. It is made of plastic and is very durable. It has exceptional clarity for viewing protozoa and is easy for students to use. This small microscope works very well when used with the plastic well slides.

The difficulty with this experiment is in viewing the tiny organisms through the small eyepiece of a microscope. It is sometimes difficult for younger students to align their eye directly into the lens so that the sample is visible. Also, these organisms can swim rapidly through the field of view, and it is easy to get frustrated trying to observe them. Methyl cellulose will help slow the organisms down without killing them. Patience with this experiment is a must. It may be useful for the students to spend one day "playing" with the microscope and observing prepared slides, pieces of hair, or other small objects before attempting to view the protozoa.

CHAPTER 6
Protists I

In the boxes provided, have the students draw and write down their observations of the movements of the protozoa.

The euglena will tend to move in a single direction, or it may not move at all, but "hover" just under the light.

A paramecium will move all over the place. It will roll, move forward and backward, and spin. There are usually other things in the water with the paramecium. Have the students note what happens when the paramecium "bumps" into other objects or other paramecia.

The amoebas move very slowly, and it can be difficult to observe them. They are usually on the bottom of the container. Allow the container to sit for 30 minutes, and then remove some solution from the very bottom, placing it on a slide. The amoebas should be visible but may be difficult to see since they are clear.

Results

Name of protozoan _____

Describe movement:

Name of protozoan _____

Describe movement:

Name of protozoan _____

Describe movement:

Draw what you observe in the pond water.

BIOLOGY LEVEL I | 57
TEACHER'S MANUAL

Have the students use their microscopes to observe the organisms in pond water or water mixed with soil. Have them draw as many organisms as they can find.

See if they can identify any protists by observing their movement and comparing this movement to the known protists they observed earlier.

CHAPTER 6
Protists I

Have the students write valid conclusions based on their observations. The conclusions will vary depending on how well the experiment worked. It is important that the students write conclusions based on their actual observations. Some examples are given.

Conclusions

We did not see any protozoa, only dirt.

We observed small protozoa that swam like paramecia.

We observed some amoebas.

We saw some organisms that we could not identify but that looked like protozoa.

Review

(Answers may vary)

Define the following terms:

protist	*a small eukaryotic creature that can be like both an animal and a plant*
microscope	*an instrument used to look at small objects*
cilia	*small "hairs" found on certain protists that help them move*
flagellum	*a long whip that is attached to many types of protists and enables a protist to move*
pseudopod	*a "false foot" found on an amoeba that allows it to move and eat*

Draw a paramecium.   Draw a euglena.   Draw an amoeba.

How do euglena and paramecia move? *Euglena and paramecia move rapidly with small hairs and whips.*

How does an amoeba move? *An amoeba moves slowly using pseudopods.*

Chapter 7: Protists II

Overall Objectives	61
7.1 Nutrition	61
7.2 How euglena eat	61
7.3 How paramecia eat	61
7.4 How amoebas eat	62
7.5 How other protozoa eat	62
7.6 Summary	62
Experiment 7: How Do They Eat?	63
Conclusions	66
Review	67

Time Required

Text reading 1 hour
Experimental 1 hour

Materials

microscope with a 10X objective[1]
microscope slides[2]
3 eye droppers
protozoa study kit from Experiment 6[3]
baker's yeast
distilled water
Eosin Y stain[4]

[1] Student Microscope, www.gravitaspublications.com, "Other Products" section, Gravitas Publications, Inc.
The following materials are available from Home Science Tools, www.hometrainingtools.com:
[2] Glass Depression Slides, MS-SLIDCON or MS-SLIDC12
[3] Basic Protozoa Set, LD-PROBASC
[4] Eosin Y, CH-EOSIN

Overall Objectives

In this chapter the students will be introduced to the feeding mechanisms of several protozoa. They will see that protists can have both plant-like and animal-like characteristics, and they will observe the feeding habits of certain protozoa.

7.1 Nutrition

Have a discussion with the students about how they get their nutrition. Then review Chapter 8 of the Chemistry Level I Student Text with them, and discuss energy molecules. Also, have a discussion about how humans are not like plants because we cannot use the sun for food.

Ask them how they think small animals like protozoa get their food.

7.2 How euglena eat

Euglena are photosynthetic protists. They have chloroplasts, just like plants, and use the sun's energy to make food by photosynthesis. Another photosynthetic protist is the Volvox which is a colonial protist — it lives in a group with other Volvox.

Euglena must be able to "detect" sunlight in order to survive. This is accomplished by a small spot near the flagellum called the eyespot or stigma. The stigma is actually a collection of photosensitive pigments that absorb photons of light. The stigma can collect light from only one direction and thus serves as a light signal to the euglena. Using the stigma, the euglena is able to orient itself in the water so that it is exposed to sufficient light for photosynthesis.

7.3 How paramecia eat

Paramecia are not photosynthetic and therefore rely on outside food sources for survival.

Paramecia have a specialized feeding structure called the oral groove. The oral groove leads to the cell mouth or cytostome, and it has cilia surrounding the entrance. The cilia beat in a circular fashion and the resulting water currents sweep the food into the oral groove.

Once inside the oral groove, the food travels to the cytostome [the cellular mouth] and from there goes into the food vacuole, which is a little sack that contains digestive enzymes. The food vacuole travels throughout the cell, digesting the food as it goes. This digestive mechanism is called intracellular digestion. The waste is excreted out a small pore called a cytoproct (cellular anus).

7.4 How amoebas eat

Amoebas eat using an entirely different mechanism than either Euglena or Paramecia. Amoebas engulf their food by using pseudopods. The amoeba moves its pseudopods in the direction of the prey, eventually engulfing it. A food vacuole is formed around the captured prey and fuses with lysosomes which inject digestive enzymes. There is no cytoproct in an amoeba, and wastes are eliminated thorough the membrane.

7.5 How other protozoa eat

Other protozoa exhibit other eating mechanisms. The illustrations in this section of the student text show examples of two protists that use different methods to capture prey. The Didinium uses a long tentacle to capture food, and the Podophrya uses its tentacles to attach itself to and remove the contents of its prey.

7.6 Summary

Discuss the main points of this chapter with the students. Explain to them that protozoa have both plant-like and animal-like qualities. Also explain that protozoa are uniquely designed and are amazing little single-celled creatures.

Experiment 7: How Do They Eat?   Date _____

Objective *In this experiment, we will observe both paramecia and amoebas eating yeast.*

Hypothesis *Different protozoa will eat in different ways.*

Materials

- microscope with a 10X objective
- microscope slides
- 3 eye droppers
- protozoa study kit from Experiment 6
- baker's yeast
- distilled water
- Eosin Y stain

Experiment

❶ Color the yeast with Eosin Y stain:

- Add one teaspoon of dried yeast to 1/2 cup of distilled water. Allow it to dissolve.
- Add one droplet of Eosin Y stain to one droplet of yeast mixture. Look at the mixture under the microscope. You should be able to see individual yeast cells that are stained red.

❷ Take the amoeba sample, and place a small droplet of the solution onto a glass slide that has been correctly positioned in the microscope.

❸ Take a small droplet of the Eosin Y stained yeast, and place it into the droplet of protozoa that is on the slide.

❹ Looking through the microscope, patiently observe the protozoa and note the red colored yeast. Try to describe how the protozoa eat. Write down as many observations as you can. Draw one of the protozoa eating.

❺ Repeat steps 2-4 using the paramecium sample.

[NOTE: This is an optional experiment. If a microscope is unavailable, or if protozoa cannot be ordered, this experiment can be skipped.]

In this experiment the students will examine how two different protozoa eat. Baker's yeast will be used as food.

Have the students read the entire experiment before writing the objective. Have the students predict whether or not the protozoa will eat the yeast. Then have them write the objective and hypothesis. Examples are given.

The Eosin Y stained yeast will be ingested by the protozoa. It may take some time for this observation. Once ingested by a protozoan, the red stained yeast will turn blue.

64 | CHAPTER 7
Protists II

Have the students write down their observations and draw a picture of an amoeba eating.

Results

Observations of an amoeba eating:

_____
_____
_____
_____
_____
_____
_____
_____
_____
_____
_____

Draw a picture showing how an amoeba eats.

Observations of a paramecium eating:
_____
_____
_____
_____
_____
_____
_____
_____
_____
_____
_____

Draw a picture showing how a paramecium eats.

BIOLOGY LEVEL I | 65
TEACHER'S MANUAL

Have the students write down their observations and draw a picture of a paramecium eating.

CHAPTER 7
Protists II

Have the students write valid conclusions based on their observations. If the experiment did not work, this should be written as a conclusion.

Conclusions

Review

(Answers may vary.)

Define the following terms:

- **stigma** — *Also called the eyespot, the stigma is used by a euglena to detect light.*
- **food vacuole** — *The sack inside a paramecium or an amoeba that holds and digest food.*
- **phagocyte** — *This term means a "cell that eats."*
- **oral groove** — *the opening in a paramecium that takes in food*
- **cytoplasm** — *the inside material of a cell*

What is a Didinium? *a protozoan that eats paramecia whole*

What is a Podophrya? *a protozoan that eats paramecia by using tentacles to suck out the insides*

Chapter 8: The Butterfly Life Cycle

Overall Objectives	69
8.1 Introduction	69
8.2 Stage I: the egg	69
8.3 Stage II: the caterpillar	70
8.4 Stage III: the chrysalis	70
8.5 Stage IV: the butterfly	70
8.6 Summary	71
Experiment 8: From Caterpillar to Butterfly	72
Conclusions	75
Review	76

Time Required

Text reading   1 hour
Experimental   about 4 weeks

Materials

locally collected caterpillar and food or butterfly kit
(suggested: Butterfly Garden, LM-BFLYGAR or butterfly pupae
LM-BFLYCUL, Home Science Tools, www.hometrainingtools.com)
small cage

Overall Objectives

In this chapter the students will examine the life cycle of the butterfly and learn about metamorphosis.

8.1 Introduction

Butterflies, moths and skippers are in the order Lepidoptera. This name reflects the fact that these insects have scaly wings. There are more than 100,000 known species, and it is the second largest insect order next to the beetles, which are in the order Coleoptera. Moths are the most abundant species in the order Lepidoptera, but butterflies are the more brightly colored and often the more familiar species.

Moths and butterflies come in a variety of shapes, sizes and coloration. The smallest moths and butterflies have wingspans that are not much larger than the size of a pencil eraser. The wingspans of the largest butterflies can extend to almost one foot.

The life cycle of Lepidoptera consists of four stages: egg, larva, pupa and adult. Some species develop in as little as three weeks. Other species can take up to three years to fully develop.

Moths and butterflies are found on every continent except Antarctica. Many species migrate from one place to another, but only the Monarch butterfly makes a true two-way migration as it travels from Mexico to North America and then back again.

8.2 Stage I: the egg

The first stage in the life cycle of butterflies is the egg. The number of eggs laid can vary from only a few hundred to several thousand. Eggs are deposited onto suitable food sources such as leaves or grasses.

Butterfly eggs come in a variety of shapes and sizes. The outer coating can be smooth and shiny, like Monarch eggs, or decorated with elaborate grooves and depressions. Some eggs are deposited as single eggs and some are deposited in groups.

The hatching of eggs coincides with favorable weather and growth of the food source. The eggs can exchange oxygen with the air via small passages in the shell, whether wet or dry.

When the egg is hatched, the larva, or caterpillar, eats the food source that the egg was attached to. Many species of moths and butterflies are limited to only a small group of suitable plants. Many species, therefore, remain in only one habitat. Other species that can eat a greater variety of foods can be found in many different habitats.

8.3 Stage II: the caterpillar

Once the egg hatches, a larvae, or caterpillar emerges. The caterpillar's sole function is to eat. This stage in the life cycle of many butterflies is the chief nutritional stage, and a caterpillar consumes many times its weight in food during this part of the cycle.

Caterpillars can be brightly colored, plain in color, hairy or smooth, depending on the species. Those caterpillars that live and feed while covered by foliage and those that are burrowers, are usually plain in color. Those caterpillars that feed in the open are usually brightly colored with ornamentation such as hair or horns that help defend against predators.

The larval stage of butterflies can last anywhere from a few weeks to several years, depending on the species. During this time, the caterpillar grows and molts, shedding the old skin as many as four or five times.

Once the caterpillar has completed the larval development stage, it stops eating and finds a suitable place to weave a chrysalis. Then the pupal stage begins.

8.4 Stage III: the chrysalis

The third stage in the butterfly life cycle is the pupal stage. The caterpillar has completed the growth and development of the larval stage and is ready for hibernation and metamorphosis.

Many species spin chrysalises on the underside of branches or other surfaces. A small bit of silk is woven at one end, and the caterpillar tests the strength of this "button" to ensure it will hold. The caterpillar then spins the silk around itself making a tough chrysalis. Many chrysalises are made of silk alone, but some species incorporate leaves, hair, or chewed wood pulp. The chrysalis sometimes has a seam that helps the adult butterfly emerge.

The pupal stage varies depending on the species. Many small species take only a few days to a few weeks to develop. Other larger species may take several months. The adult will emerge only when conditions are right, and some chrysalises have been known to survive several years before the adult finally comes out.

8.5 Stage IV: the butterfly

The final stage in the butterfly life cycle is the adult stage. Once the chrysalis has completed the time required for metamorphosis, and when conditions permit, the adult butterfly will emerge from the chrysalis.

To exit the chrysalis, the fully formed adult butterfly wriggles until it is finally free. Some species have spines along their back to help bore holes in the chrysalis and help push out the walls.

When the young butterfly emerges, it cannot yet fly. Since the wings are wrinkled, the butterfly must pump fluid into them to flatten and stiffen them. Often the butterfly crawls to a place where it can hang with its head up, allowing the fluid to flow into the wings. It may take several minutes or even a few hours before the wings are stiff enough for flight.

The main purpose of the adult stage in the life cycle of butterflies and moths is reproduction. Nutrition is essential only in a few species during this stage. Food is taken in only for supplying energy for flight.

Many butterflies and moths travel great distances during the adult stage. In North America many moths migrate to Canada, and in Europe, many butterflies and moths migrate to Scandinavia. Once a butterfly has found a mate, the female lays eggs and the cycle repeats.

8.6 Summary

Discuss the summary statements for this chapter.

CHAPTER 8
The Butterfly Life Cycle

In this experiment, the students will observe the change from caterpillar to butterfly. There is no hypothesis for this experiment. A sample objective is given.

If possible, have the students collect local caterpillars and house them in a small cage. If a caterpillar can be located, take several leaves from the plant where it was found so the caterpillar will have food. If a local caterpillar cannot be found, caterpillar kits can be purchased.

Resources:

Home Science Tools sells a butterfly kit:
Butterfly Garden, LM-BFLYGAR
www.hometrainingtools.com

They also sell butterfly pupae if you would rather get just the caterpillars instead of a kit. If you choose to buy the butterfly pupae, you will be sent a coupon that you can then redeem online or in the mail for delivery of the pupae within 2-3 weeks after the coupon is redeemed.
Butterfly Pupae, LM-BFLYCUL
www.hometrainingtools.com

Experiment 8: From Caterpillar to Butterfly   Date _____

Objective *We will observe the change (metamorphosis) as a caterpillar turns into a butterfly.*

Materials
locally collected caterpillar and food, or butterfly kit
small cage

Experiment
1. Place your local caterpillar in a small cage and provide food for it, or follow the directions on the butterfly kit for proper care of your caterpillar.
2. Fill out the life cycle chart on the next page.
3. Over the course of the next several weeks, observe any changes your caterpillar undergoes.
4. Record how much food your caterpillar eats.
5. Record how many times the caterpillar molts.
6. Record where the caterpillar spins its chrysalis.
7. If you can observe the caterpillar emerging as a butterfly, record how long it takes before it can fly.

Draw the various stages in the life cycle of a butterfly.

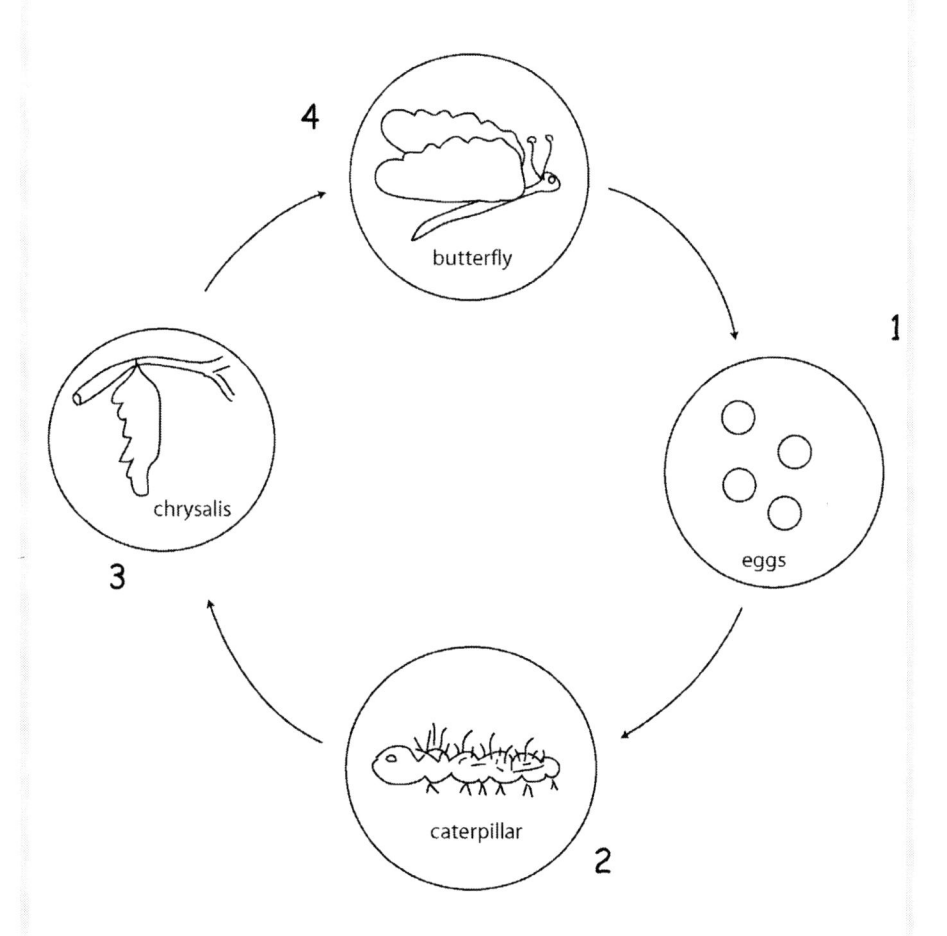

4 butterfly

1 eggs

2 caterpillar

3 chrysalis

74 | CHAPTER 8
The Butterfly Life Cycle

Have the students write their observations in the Results section. It may not be possible to determine how much the caterpillar eats or all of the molting stages. Have the students record any other observations, such as movement or periods of inactivity.

Results

Week	Amount of food eaten	Molting?	Other observations
1			
2			
3			
4			
5			
6			
7			
8			
9			
10			
11			
12			

Conclusions

BIOLOGY LEVEL I | 75
TEACHER'S MANUAL

Have the students write some valid conclusions based on their observations. Some examples are:

- *I observed the caterpillar eating only three full leaves.*
- *I recorded only three molts for the caterpillar.*
- *It took two weeks for the caterpillar to form a chrysalis.*

Again, help the students be accurate with their concluding statements.

CHAPTER 8
The Butterfly Life Cycle

Review

(Answers may vary.)

Define the following terms:

Lepidoptera	*This term means "wings with scales."*
larval stage	*the caterpillar stage of the butterfly life cycle*
molting	*when a caterpillar or other animal sheds its skin as it grows*
pupal stage	*the chrysalis stage of the butterfly life cycle*
chrysalis	*the casing a caterpillar creates when it's ready to turn into a butterfly*
imago	*the fully developed insect*

Chapter 9: The Frog Life Cycle

Overall Objectives	78
9.1 Introduction	78
9.2 Stage I: the egg	78
9.3 Stage II: the tadpole	78
9.4 Stage III: from tadpole to frog	79
9.5 Stage IV: the adult frog	79
9.6 Summary	79
Experiment 9: From Tadpole to Frog	80
Conclusions	84
Review	85

Time Required

    Text reading   1 hour
    Experimental   about 4 weeks

Materials

    tadpoles[1]
    tadpole food[1]
    small aquarium
    tap water conditioner and tap water OR distilled water

[1] Grow-A-Frog Kit, M-GROFROG, Home Science Tools, www.hometrainingtools.com

Overall Objectives

In this chapter the students will be introduced to the life cycle of frogs. They will examine the various stages of the life cycle and observe a tadpole changing into an adult frog.

9.1 Introduction

Frogs are amphibians and are in the class Amphibia. The term amphibia means "both lives" and refers to those creatures that live both in water and on land for at least part of their lives. However, not all amphibians live their lives both in the water and on land. Some spend their whole lives either on land or in the water, but even those that live almost exclusively on land need to be near water in a moist environment.

Frogs are not the only amphibians. The class Amphibia also includes salamanders, newts, mud puppies, sirens, and caecilians. There are about 5,000 known species of amphibians. Frogs are in the order Anura, which means "without tail," since frogs lose their tails as adults. Salamanders do not lose their tails and are in the order Caudata, which means "having a tail."

9.2 Stage I: the egg

All amphibians begin life as an egg. The amount of time spent as an egg varies from a few hours to a few weeks. The outside of the egg does not have a tough outer shell, like a chicken egg does. Instead, a jelly-like coating covers the outside of the egg. Often the eggs are laid in clumps or in long strings. The jelly-like coating protects the eggs from drying out and helps them stick together.

The maternal parent will often attend the eggs until the time they hatch. Some frogs, such as the Pipa carry the eggs and young on their backs, and the recently extinct Rheobatrachus females swallowed their eggs and incubated their young in their stomachs!

9.3 Stage II: the tadpole

When the eggs hatch, tadpoles, or pollywogs, emerge. The tadpole stage is considered the larval stage since the tadpoles are immature amphibians. Tadpoles differ from their adult counterparts in significant ways. They have tails and gills and no eyelids.

The size of a tadpole varies depending on the species. Often, tadpoles are larger than the adults. In general, tadpole size increases with length of the developmental period: smaller tadpoles have shorter developmental stages and larger tadpoles have longer developmental stages. The Harlequin tadpole grows up to 10 inches long and takes four months to develop. However, the adult frog is only 3 inches long. This reduction in size is called "shrinkage."

9.4 Stage III: from tadpole to frog

The process of changing from a juvenile tadpole into an adult frog is called metamorphosis. The many changes that occur in frogs during metamorphosis include the emergence of limbs and the absorption of the tail into the body. The skin thickens and lungs develop as the external gills disappear. There are also dramatic changes that occur in the digestive tract that are associated with a change in diet.

Other amphibians, such as salamanders, do not undergo such dramatic changes as do frogs.

9.5 Stage IV: the adult frog

There are many different species of frogs, and they make up the biggest order of amphibians. There are true frogs, tree frogs, tropical frogs, toads, narrow mouth toads, and spade foot toads, among others. All of these species differ slightly in their coloring and characteristics, but they do have some similar features.

Most species of frogs have well-developed hearing. They have a membrane called the tympanic membrane or tympanum that is located just behind the eyes. This membrane vibrates in response to sound.

Most frogs have eyes that see very well. Land frogs have an eyelid, and water frogs have a thin membrane covering the eye for protection. Eye color and shape varies, and most frogs have horizontal pupils.

Frogs come in a variety of colors. Many poisonous frogs are brightly colored, warning potential predators that they are poisonous.

9.6 Summary

Review the summary statements at the end of the chapter.

CHAPTER 9
The Frog Life Cycle

In this experiment, the students will observe the life cycle of a frog.

Have the students read the experiment and then write an objective. There is no hypothesis for this experiment.

If possible, observe the growth and development of a frog from a tadpole collected from a local pond or stream. However, if this is not possible, tadpoles and/or frog eggs can be purchased.

Resources:

Two sources of instructions for caring for tadpoles that you collect locally are:

www.allaboutfrogs.org/info/tadpoles/index.html

www.live-tadpoles.com/raising-tadpoles.html

More information can be found online.

A tadpole kit can be ordered from Home Science Tools. The tadpole provided is transparent, so the internal organs can be easily seen. Also, it is fairly hearty and can survive a few skipped meals.

Grow-A-Frog Kit, M-GROFROG
www.hometrainingtools.com

NOTE: The Xenopus frog is NOT recommended.

Experiment 9: From Tadpole to Frog   Date: _____

Objective   *We will observe the change (metamorphosis) as a tadpole turns into a frog*

Materials

tadpole (or tadpole kit)
tadpole food
small aquarium
tap water conditioner and tap water (or distilled water)

Experiment

❶ "Cure" the tap water by adding tap water conditioner. This removes any chemicals, like chlorine, that would be harmful to the tadpoles and frogs.

❷ Fill the aquarium 1/2 to 3/4 full with the conditioned water.

❸ Add the live tadpoles.

❹ Feed the tadpoles according to the directions.

❺ Observe the changes the tadpole makes over the course of 4 to 6 weeks.

❻ Record your observations in the Results section. See if you can identify the different stages of the tadpole that are outlined in this chapter of the student text. (Note when the legs emerge, note the time it takes for the front legs to emerge etc.)

Draw the various stages in the life cycle of a frog.

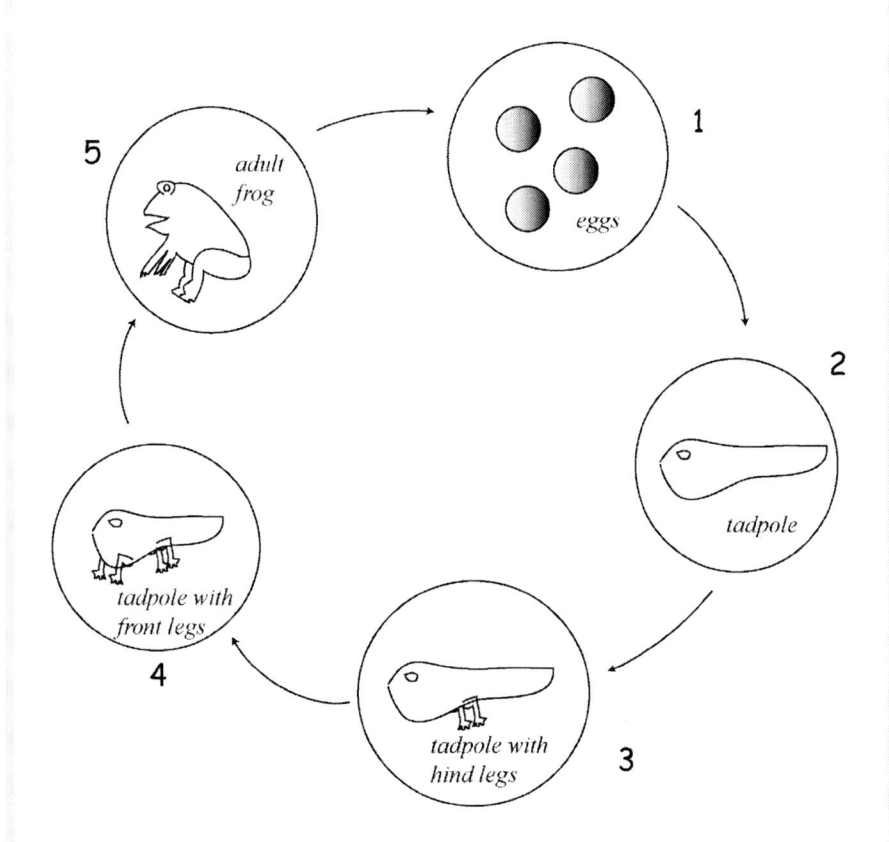

1. eggs
2. tadpole
3. tadpole with hind legs
4. tadpole with front legs
5. adult frog

82 | CHAPTER 9
The Frog Life Cycle

Label the parts of the frog.

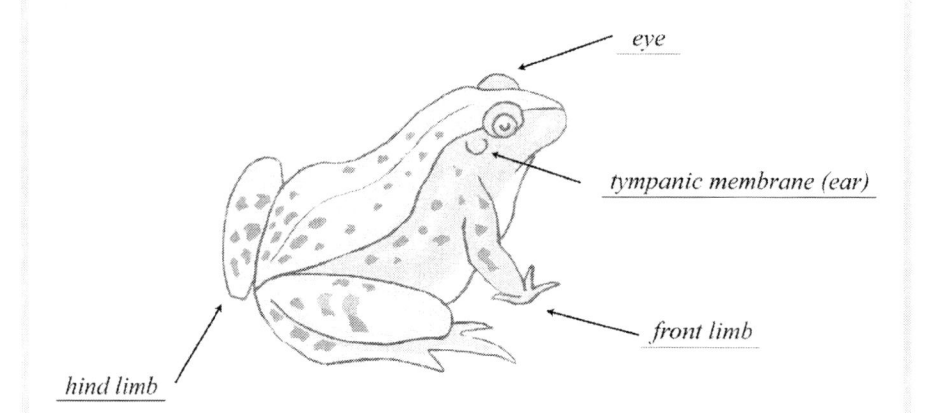

eye

tympanic membrane (ear)

front limb

hind limb

Results

Week

1 _____

2 _____

3 _____

4 _____

5 _____

6 _____

7 _____

8 _____

9 _____

10 _____

11 _____

12 _____

BIOLOGY LEVEL I | 83
TEACHER'S MANUAL

Have the students record the changes they observe over the course of several weeks.

Have them note in particular if the hind legs do indeed emerge before the front legs. Also have them observe whether the hind legs begin as buds and the front legs come out fully developed.

CHAPTER 9
The Frog Life Cycle

Have the students write some valid conclusions based on their observations.

Conclusions

BIOLOGY LEVEL I | 85
TEACHER'S MANUAL

Review

(Answers may vary.)

Define the following terms:

spawning — *when an adult female frog lays eggs to be fertilized by the male frog*

amphibian — *a term that means "both lives"*

metamorphosis — *for amphibians, a term that refers to the change from tadpole to frog*

tympanic membrane — *the ear of a frog*

Chapter 10: Our Balanced World

Overall Objectives	87
10.1 Introduction	87
10.2 Ecosystems	87
10.3 The food cycle	88
10.4 The air cycle	88
10.5 The water cycle	88
10.6 Summary	89
Experiment 10: Making an Ecosystem	90
Conclusions	92
Review	93

Time Required

Text reading   30 minutes
Experimental   about 3-5 weeks

Materials

clear glass or plastic tank with a solid lid
water
plastic wrap (e.g., Saran Wrap)
soil
small plants
small bugs:
   worms
   small beetles
   ants, etc.

Overall Objectives

In this chapter the students will be introduced to the concept of an ecosystem and the often delicate balance that sustains an ecosystem.

10.1 Introduction

The Earth can be considered a global ecosystem — an ecosystem that constitutes the entire globe. Within this global ecosystem, smaller ecosystems are found. These include everything from a small pond to large lakes, oceans, and other habitats.

10.2 Ecosystems

Have a discussion with the students about the balance that exists between plants, animals, and other organisms on the planet. Have the students think about the plants and animals they know about or have found in their backyard or local environment. Ask the students what is required for plants and animals to live. Lead a discussion with the following questions:

- What do plants need to have in order to live? *(light, air, water)*
  Where do plants get their food? *(photosynthesis, soil, air)*
  What do they need for food? *(carbon dioxide, water, nitrogen, minerals)*

- What do animals need to have in order to live? *(food, air, water)*
  Where do animals get their food? *(plants, other animals)*

- What do we need to have in order to live? *(food, water, air, shelter, clothing)*
  Where do we get our food? *(plants, animals)*
  Where do we get our shelter and clothing? *(plants, animals, non-living things like rocks)*

- What would happen to the plants if there were no water?

- What would happen to the animals if there were no plants?

- What would happen to us if there were no plants or animals?

Have a discussion with the students about the meaning of a cycle. A cycle is simply a sequence of events that repeats itself. Ask the students what cycles they have observed. They should reply that they observed the life cycles of frogs and butterflies in previous experiments.

Have a discussion with the students about the difference between the life cycle of a living creature and the cycles involved in non-living phenomena, such as weather. The life cycle of a living creature depends on the survival of the creature. Once the animal or plant is extinct, or no longer surviving, the cycle ceases. A weather cycle, on the other hand, continues. The weather may vary, and the cycle may change patterns, but it continues unless the Earth itself disappears.

10.3 The food cycle

Discuss the food cycle of living things. All living things require some energy source for survival. The food cycles of living things are interrelated. Microorganisms use the remains of decaying animals for food. These microorganisms provide nutrients for plants. Plants are the primary food source of many animals, and some animals feed directly on other animals and do not eat plants.

Point out the connections between various animals and their food sources. Explain how each one is dependent on the other for survival.

10.4 The air cycle

Have a discussion with the students about how we get the air we breathe. Explain that air is also in a cycle. We inhale oxygen and exhale carbon dioxide. Carbon dioxide is a gas, just as oxygen is. As part of the cycle, plants use carbon dioxide to make food for photosynthesis. Then they give off oxygen. Thus, carbon dioxide and oxygen are constantly recycled. Ask the students what would happen if animals used carbon dioxide and gave off oxygen as do plants (where would we get more carbon dioxide?).

Have a discussion about how this balance between carbon dioxide and oxygen helps keep living things alive and how plants and animals are interdependent.

10.5 The water cycle

Have a discussion about the way water is cycled from place to place. Evaporation of water from the oceans forms clouds made of very fine water droplets. When the clouds travel over land, they deposit their water in the form of rain. The rain collects in rivers, streams and underground aquifers, providing water for many living things. Small streams feed larger streams which feed rivers. Rivers empty their water back into the ocean and the cycle repeats.

Water is vital for living creatures, and without water, living things die. Some living things can go without water for many days, even weeks, but no living thing can survive without some water.

The weather cycles are driven by solar radiation, the spinning of the Earth on its axis, and the tilt of the Earth. These are balanced to provide the variety of winds, rain, snow, and hot weather that occur on the globe. The atmosphere also insulates the earth, providing the temperatures needed for life. The atmosphere, although transparent, traps heat close to the earth's surface, keeping a moderately warm average temperature.

There are other example of cycles that provide balance to our ecosystem. Nitrogen, phosphorus, sulfur, and carbon are all cycled throughout the globe.

10.6 Summary

Discuss the points in the Summary section in the student text. Have a discussion about how all of the cycles covered in the text are interrelated. Bodies of water produce the clouds that provide water for the plants and animals. The plants provide oxygen and food for the animals. The animals provide carbon dioxide for plants and food for other animals and small organisms, and so on.

CHAPTER 10
Our Balanced World

In this experiment the students will build a small ecosystem.

Have the students read the experiment and write an objective. An example is given.

In the first part of the experiment the students will observe the evaporation and condensation of water. In Part I A the water will not escape the container, but will instead condense on the top or sides. The condensed water, will "rain" back to the bottom. Explain that this is how water cycles in our atmosphere. In Part I B, the lid is left off the tank. Explain how in this case, the water disappears altogether and is not cycled back into the ecosystem in the container. Explain that if we did not have an atmosphere, the water on our planet would disappear.

Closed ecosystems are difficult to maintain. The food sources, light, and mineral nutrients need to be properly balanced. However, it is recommended that a closed ecosystem be attempted so that the students can observe the changes and problems that occur.

For the closed ecosystem in this experiment, use a few plants and some small animals like bugs and worms. These living things may die, but it is possible that the closed system will work long enough to observe the interactions of the various organisms.

Choose only one or two plants and a few small bugs. It is not necessary to have many plants and animals, and this would only complicate the experiment. Don't use frogs, lizards or other larger animals. Anticipate the food sources necessary for the plants and animals you choose.

Experiment 10: Making an Ecosystem   Date:_____

Objective   *In this experiment we will try to build an ecosystem and will observe what happens over the course of several weeks.*

Materials

- clear glass or plastic tank with a solid lid
- water
- plastic wrap
- soil
- small plants
- small bugs such as worms, small beetles, ants, etc.

Experiment

1. Take the glass or plastic tank, and cover the bottom with water.
2. Cover the tank securely with plastic wrap, and allow it to sit overnight.
3. Record your observations in Part I A of the Results section.
4. Remove the plastic wrap and let it sit overnight again.
5. Record your results in Part I B of the Results section.
6. Place the soil on the bottom of the container. Put in enough soil to fill the container about 1/3 full. Add some water if needed.
7. Plant the small plants in the soil.
8. Add the small bugs to the tank.
9. Place the lid on the tank. This is now a small closed ecosystem.
10. In Part II of the Results section, record any changes in your ecosystem that occur over the next several weeks.

Results

Part I

A _____
  _____

B _____
  _____

Part II

BIOLOGY LEVEL I | 91
TEACHER'S MANUAL

Use enriched soil. This can be purchased, or you can use soil formed with compost. This soil contains the necessary microbes for the ecosystem. Assemble the plants, bugs, and soil and add a little water. Don't add too much water. Seal the container and the closed ecosystem is complete. Place the ecosystem in indirect sunlight.

Have the students record their observations in the Results section. For Part I A and Part I B have them write down what happens to the water.

For Part II, their observations can be written or sketched. The observations will vary depending on what happens to the ecosystem. If too much water is added, the plants may begin to mold or die. If there is not enough food or water for the worms or bugs, they may die. The leaves of the plants may change color if there is not enough nitrogen or other mineral nutrients in the soil.

CHAPTER 10
Our Balanced World

Have the students write conclusions based on their observations. If the ecosystem developed problems, have the students try to guess the causes: too much or too little water, too much light, too little light, not enough food, etc

Discuss with the students how difficult it is to maintain a balanced ecosystem and how remarkable our ecosystem, the Earth, is. All of the living and non-living systems cycle and are balanced in just the right amounts needed to provide a continuous environment for future generations of living things. Have a discussion about the orbital path of the Earth being elliptical, providing the necessary seasons, and how the spinning of the earth causes wind and other weather needed for living things to survive. Discuss how the sun is just far enough away that it is not too hot, but close enough to provide adequate heat and light. All of these factors contribute to make our unique world.

Closed ecosystems can also be purchased. NASA has developed the Ecosphere—a glass sphere containing an enclosed ecosystem. It comes in spheres of various sizes and can be purchased at some nature stores or online. The Ecospheres are completely enclosed and can survive for many years.

Open ecosystems, or terraria, can also be made. A terrarium can be easily assembled and can include frogs, lizards, or salamanders. Fish tanks are also a kind of open ecosystem. The plants and animals in terraria need an outside source of food and water, and the tanks need to be cleaned, but they have a higher survival rate than closed systems.

Conclusions

_____
_____
_____
_____
_____
_____
_____
_____
_____
_____
_____
_____
_____
_____
_____
_____
_____
_____
_____
_____
_____
_____

Review

(Answers may vary.)

Define the following terms:

ecosystem	*a community of plants, animals and other creatures living together*
cycle	*a series of steps that repeats continuously*
food cycle	*the steps are: plants feed animals, animals feed other animals, animals die and feed microorganisms, microorganisms feed plants*
air cycle	*the steps are: animals breathe oxygen and exhale carbon dioxide, plants take in carbon dioxide and give off oxygen*
water cycle	*the steps are: rain from clouds falls onto land, feeding streams and rivers; rivers feed oceans; oceans supply water to make clouds*

CHAPTER 10
Our Balanced World